T0155566

SpringerBriefs in Molecular Science

Chemistry of Foods

Series editor

Salvatore Parisi, Industrial Consultant, Palermo, Italy

The series Springer Briefs in Molecular Science: Chemistry of Foods presents compact topical volumes in the area of food chemistry. The series has a clear focus on the chemistry and chemical aspects of foods, topics such as the physics or biology of foods are not part of its scope. The Briefs volumes in the series aim at presenting chemical background information or an introduction and clear-cut overview on the chemistry related to specific topics in this area. Typical topics thus include:

- Compound classes in foods—their chemistry and properties with respect to the foods (e.g. sugars, proteins, fats, minerals, …)
- Contaminants and additives in foods—their chemistry and chemical transformations
- Chemical analysis and monitoring of foods
- Chemical transformations in foods, evolution and alterations of chemicals in foods, interactions between food and its packaging materials, chemical aspects of the food production processes
- Chemistry and the food industry—from safety protocols to modern food production

The treated subjects will particularly appeal to professionals and researchers concerned with food chemistry. Many volume topics address professionals and current problems in the food industry, but will also be interesting for readers generally concerned with the chemistry of foods. With the unique format and character of Springer Briefs (50 to 125 pages), the volumes are compact and easily digestible. Briefs allow authors to present their ideas and readers to absorb them with minimal time investment. Briefs will be published as part of Springer's eBook collection, with millions of users worldwide. In addition, Briefs will be available for individual print and electronic purchase. Briefs are characterized by fast, global electronic dissemination, standard publishing contracts, easy-to-use manuscript preparation and formatting guidelines, and expedited production schedules. Both solicited and unsolicited manuscripts focusing on food chemistry are considered for publication in this series.

More information about this series at http://www.springer.com/series/11853

Dongliang Ruan · Hui Wang
Faliang Cheng

The Maillard Reaction in Food Chemistry

Current Technology and Applications

 Springer

Dongliang Ruan
Dongguan University of Technology
Dongguan, Guangdong, China

Faliang Cheng
Dongguan University of Technology
Dongguan, Guangdong, China

Hui Wang
Lacombe Research Centre
Agriculture and Agri-Food Canada
Lacombe, AB, Canada

ISSN 2191-5407   ISSN 2191-5415   (electronic)
SpringerBriefs in Molecular Science
ISSN 2199-689X   ISSN 2199-7209   (electronic)
Chemistry of Foods
ISBN 978-3-030-04776-4   ISBN 978-3-030-04777-1   (eBook)
https://doi.org/10.1007/978-3-030-04777-1

Library of Congress Control Number: 2018962133

© The Author(s), under exclusive license to Springer Nature Switzerland AG 2018
This work is subject to copyright. All rights are reserved by the Publisher, whether the whole or part of the material is concerned, specifically the rights of translation, reprinting, reuse of illustrations, recitation, broadcasting, reproduction on microfilms or in any other physical way, and transmission or information storage and retrieval, electronic adaptation, computer software, or by similar or dissimilar methodology now known or hereafter developed.
The use of general descriptive names, registered names, trademarks, service marks, etc. in this publication does not imply, even in the absence of a specific statement, that such names are exempt from the relevant protective laws and regulations and therefore free for general use.
The publisher, the authors, and the editors are safe to assume that the advice and information in this book are believed to be true and accurate at the date of publication. Neither the publisher nor the authors or the editors give a warranty, express or implied, with respect to the material contained herein or for any errors or omissions that may have been made. The publisher remains neutral with regard to jurisdictional claims in published maps and institutional affiliations.

This Springer imprint is published by the registered company Springer Nature Switzerland AG
The registered company address is: Gewerbestrasse 11, 6330 Cham, Switzerland

Preface

Maillard reaction, the well-known non-enzymatic reaction between reducing sugars and proteins, is one of the most important reactions in food sciences. Amino acids, peptides and proteins are the key constituents of food and also directly contribute to the changes in colour, flavor, and nutritive value but also for the formation of stabilizing and mutagenic compounds during thermal or enzymatic reactions in production, processing and storage of food. Proteins also contribute significantly to the physical properties of food through their ability to build or stabilize gels, foams and emulsions properties. Due to the complexity of the Maillard reaction, mass spectrometry (MS) is a useful and powerful technology to characterize the Maillard reaction products and we have apply MS-based technologies to systematically investigate Maillard reaction from amino acids, peptides, proteins, to characterize the Maillard reaction products and to identify the modification sites of proteins. Studies demonstrate that glucosylation increased emulsification activity index and emulsifying stability index. The present investigation demonstrates that MS-related technologies are important tools to analyze and characterize the Maillard reaction products and functional properties of food proteins.

Dongguan, China   Dongliang Ruan
Lacombe, Canada   Hui Wang
Dongguan, China   Faliang Cheng

Acknowledgements

The authors would like to thank Prof. C. Y. Ma and Dr. Ivan K. Chu for their invaluable advices, constant encouragement and inspiring discussion during the studies and writing process. The authors would like to thank Prof. Michael Siu, Dr. Wang Mingfu, Dr. K. H. Sze and Dr. Huang Junchao for their stimulating discussion and helpful comments in the research.

The authors would like to thank my friends, colleagues and members of our laboratories for reading and comments on chapters: Dr. Siu Shiu On, Dr. Emily Choi, Dr. Song Tao, Dr. Edward Lau and Dr. Corey N. W. Lam for the valuable friendship, help and support.

The authors would like to express my appreciation to all the staff and technical service crew in the School of Biological Science and the Chemistry Department, the University of Hong Kong.

Finally, the authors would like to dedicate the whole-heart thanks to Hui, Irene and Max, and all the family members for their love, care and understanding throughout my research.

Contents

Abbreviations

12-crown-4	1,4,7,10-tetraacyclododecane
ACN	Acetonitrile
CID	Collision induced dissociation
ECD	Electron capture dissociation
EI	Electron impact
ESI	Electrospray
ESI*	Emulsifying stability index
ETD	Electron transfer dissociation
eV	Electron voltage
FAB	Fast atom bombardment
FTICR	Fourier transform ion cyclotron resonance
GC	Gas chromatography
HPLC	High pressure/performance liquid chromatography
HRP	Horseradish peroxidase
LIT	Linear ion trap
m/z	Mass to charge ratio
MALDI	Matrix-assisted laser desorption/ionization
MeOH	Methanol
MS	Mass spectrometry
MS^2 or MS/MS	Tandem mass spectrometry
MS^n	Multistage tandem mass spectrometry
NMR	Nuclear magnetic resonance
NP	Normal phase
PMF	Peptide mass fingerprint
PNGaseF	Peptide N-glycosidase F
PTMs	Posttranslational modifications
Q	Quadrupole
QIT	Quadrupole ion trap
QqQ	Triple quadrupole
RNaseB	Ribonuclease B

RP	Reversed phase
SE	Single extraction
Terpy	2,2':6',2″ –terpyridine
TFA	Trifluoro acetic acid
TOF	Time-of-flight
UPLC	Ulta-high pressure/performance liquid chromatography
XIC	Extracted ion chromatogram

List of Figures

List of Tables

List of 20 Common Amino Acids

Name	Abbr	Structure	Monoisotopic amino acid residue mass
Glycine	Gly/G	$$H_2N-\underset{\underset{H}{\mid}}{CH}-\overset{\overset{O}{\parallel}}{C}-OH$$	57.02146
Alanine	Ala/A	$$H_2N-\underset{\underset{CH_3}{\mid}}{CH}-\overset{\overset{O}{\parallel}}{C}-OH$$	71.03711
Isoleucine	Ile/I	$$H_2N-\underset{\underset{\underset{\underset{CH_3}{\mid}}{CH_2}}{\underset{\mid}{CH-CH_3}}}{CH}-\overset{\overset{O}{\parallel}}{C}-OH$$	113.08406
Leucine	Leu/L	$$H_2N-\underset{\underset{\underset{\underset{CH_3}{\mid}}{CH-CH_3}}{\underset{\mid}{CH_2}}}{CH}-\overset{\overset{O}{\parallel}}{C}-OH$$	113.08406

(continued)

(continued)

Name	Abbr	Structure	Monoisotopic amino acid residue mass
Proline	Pro/P	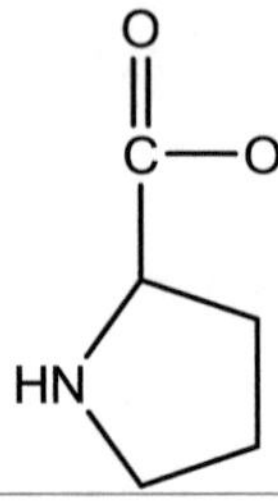	97.05279
Valine	Val/V	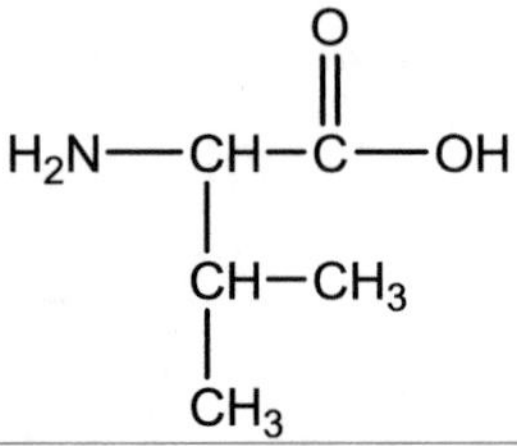	99.06841
Aspartic acid	Asp/D	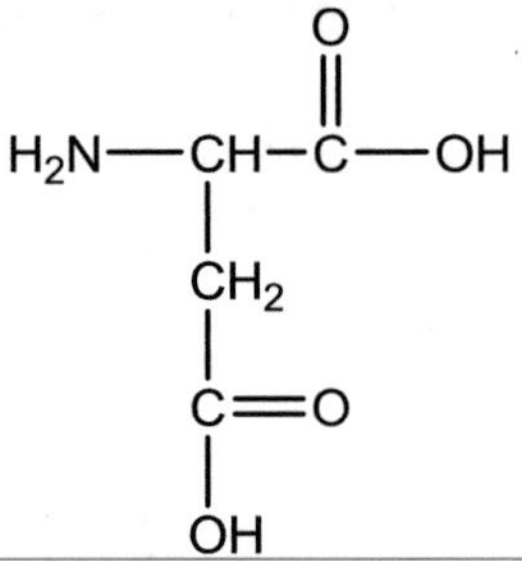	115.02694
Glutamic acid	Glu/E	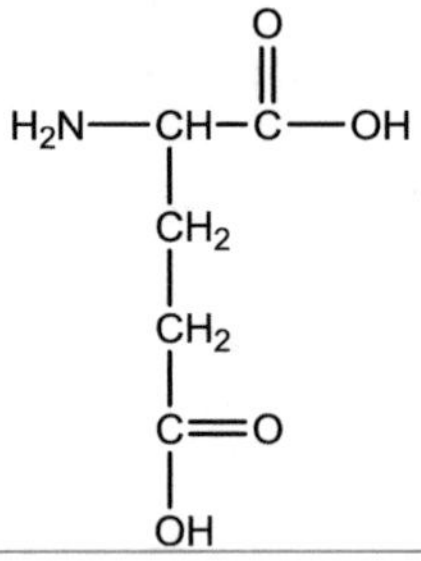	128.04259

(continued)

(continued)

Name	Abbr	Structure	Monoisotopic amino acid residue mass
Arginine	Arg/R	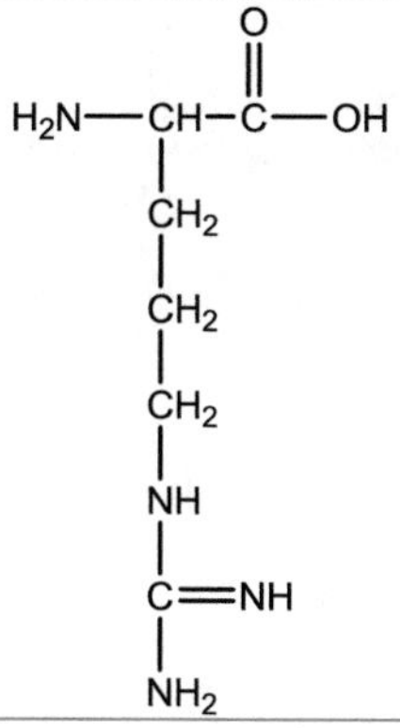	156.10111
Histidine	His/H	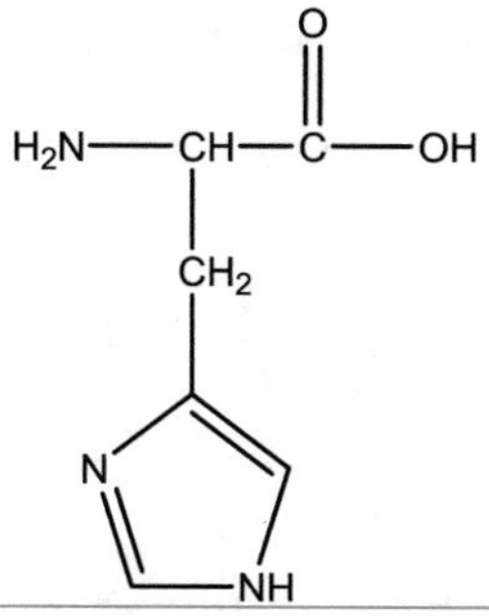	137.05891
Lysine	Lys/K	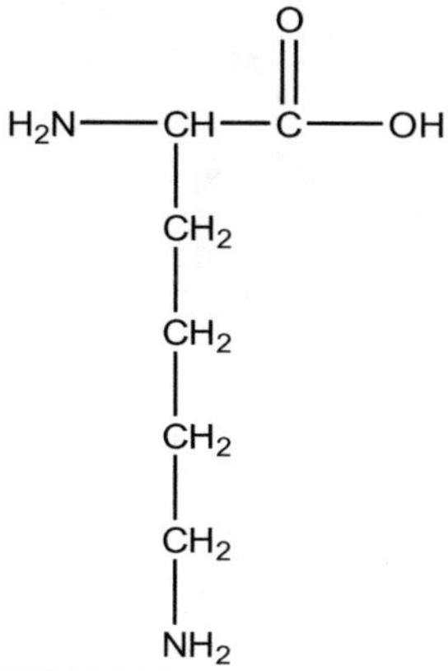	128.09497

Arginine structure:

$$H_2N-CH-C(=O)-OH$$
with side chain: $-CH_2-CH_2-CH_2-NH-C(=NH)-NH_2$

Histidine structure:

$$H_2N-CH-C(=O)-OH$$
with side chain: $-CH_2-$ (imidazole ring, N...NH)

Lysine structure:

$$H_2N-CH-C(=O)-OH$$
with side chain: $-CH_2-CH_2-CH_2-CH_2-NH_2$

(continued)

(continued)

Name	Abbr	Structure	Monoisotopic amino acid residue mass
Phenylalanine	Phe/F	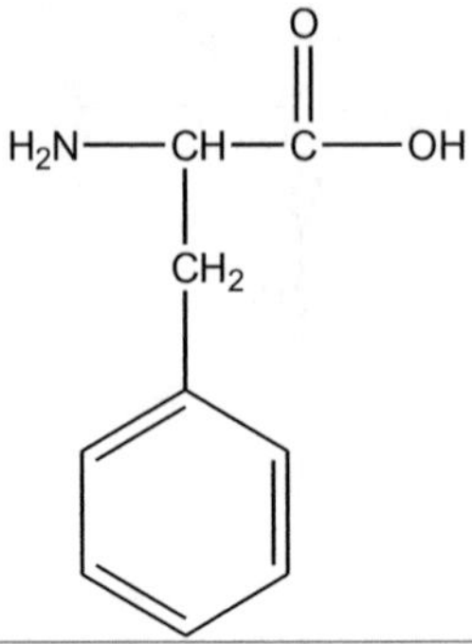	147.06842
Tryptophan	Trp/W	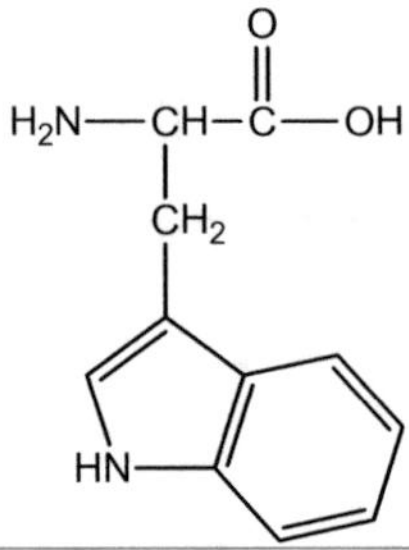	186.07931
Tyrosine	Tyr/Y	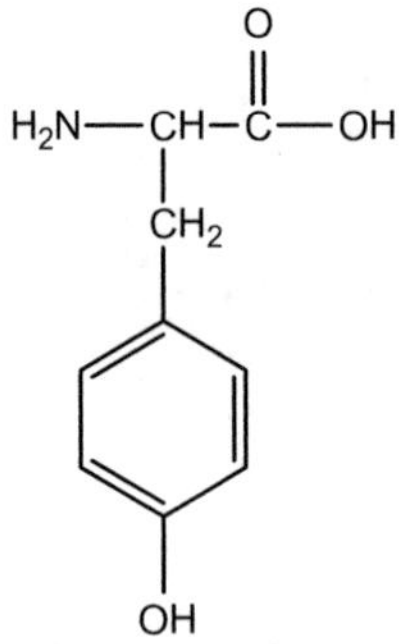	163.06333
Asparagine	Asn/N	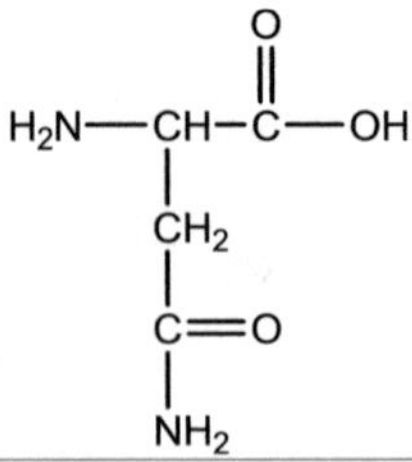	114.04293

(continued)

(continued)

Name	Abbr	Structure	Monoisotopic amino acid residue mass
Cysteine	Cys/C	$H_2N-CH-C(=O)-OH$, with side chain CH_2-SH	103.00919
Glutamine	Gln/Q	$H_2N-CH-C(=O)-OH$, with side chain $CH_2-CH_2-C(=O)-NH_2$	128.05858
Methionine	Met/M	$H_2N-CH-C(=O)-OH$, with side chain $CH_2-CH_2-S-CH_3$	131.04049
Serine	Ser/S	$H_2N-CH-C(=O)-OH$, with side chain CH_2-OH	87.03203
Threonine	Thr/T	$H_2N-CH-C(=O)-OH$, with side chain $CH-OH$ and CH_3	101.04768

Chapter 1
The Maillard Reaction

1.1 Introduction

The Maillard reaction, named after the French chemist Louis-Camille Maillard who discovered it in 1912, is a non-enzymatic reaction that takes place between an available amino group and a carbonyl-containing moiety [1, 2]. The Maillard reaction spontaneously takes place in food in the presence of heat, wherein the reactive carbonyl groups of reducing sugars react with the nucleophilic amino group of amino acids, peptides, or proteins to form a large variety of compounds. These Malliard products are often the determinant elements for the flavors, odors, functionality, and nutrition of food [1–4]. In addition, extensive protein modification and cross-linking occur, resulting in changes in emulsifying properties, thermal stability, and water-holding capacity of proteins [5–9].

As food products are commonly rich in proteins and carbohydrates, Maillard reactions promise a powerful way to alter the functional properties of food [6, 10–12]. Efforts from food and chemical scientists to understand Maillard reaction range from studying reactive behaviors in simple amino acid–carbohydrate models to complex protein–sugar systems. Many studies focus on the identification of reaction-available lysine residues, characterization of Maillard product structure, and the evaluation of the impact of processing on the nutritional quality of food proteins [5, 13, 14].

In particular, a great amount of studies exists on the beneficial functionality of mild degrees of Maillard glycoconjugates. Protein glycation during the early stages of Maillard reaction results in benefits such as increases in emulsification and calcium complexing activity [15], improved foaming properties [16] and protein solubility [17], enhanced antimicrobial activity, bactericidal activity, and heat stability [18–20]. Beneficial effects have also been observed at more advanced stages of the Maillard reaction, including the formation of compounds with antioxidant, anticarcinogenic, and antimutagenic properties [21]. The cross-linking of proteins has also been associated with improved food texture [22]. However, these studies

© The Author(s), under exclusive license to Springer Nature Switzerland AG 2018
D. Ruan et al., *The Maillard Reaction in Food Chemistry*, Chemistry of Foods
https://doi.org/10.1007/978-3-030-04777-1_1

could not demonstrate clearly the reaction conditions that affect the extent of Maillard reaction at the molecular level due to poor monitoring and control of the reaction. Furthermore, due to the limitation of the traditional spectrometric methods used (e.g., UV–vis and fluorescence spectroscopy), limited evidences are available to illustrate the mechanism of glycation between sugars and proteins in detail. Potentially important information for food modification such as changes in food properties with different number of sugar moieties is also missing.

1.2   The Maillard Reaction and Food Properties

One of the major practical impacts of the Maillard reaction is the formation of different kinds of flavors and aromas in food. Through heating, boiling, cooking, or storage, Maillard products transform a great variety of food and raw ingredients into palatable dishes with desirable and diversified aromas and flavors [23–25]. As our knowledge on the Maillard reaction progresses with the development of new analytical technologies, many significant flavor products have been observed. By correlating their aromas, structures, molecular shapes with their routes of formation, Maillard reaction flavors can be classified into four main groups including nitrogen heterocyclics, cyclic enolones, monocarbonyls, and polycarbonyls [26–28].

Although the Maillard reaction serves to improve food quality, it is also associated with losses of nutritive value such as damage to basic amino acids and vitamins, impairment of protein digestion, and formation of toxic compounds [29, 30]. Among these negative consequences, loss of food quality and possible decrease of food safety have drawn the most attention [5, 28, 31]. For example, since the Maillard reaction takes place under conditions of high temperature and low moisture, the essential amino acids of food proteins with free terminal amines, particularly lysine, can react with sugars to form enzyme-resistant cross-linkages, resulting in irreversible protein damage and decrease of protein availability and digestibility [32–34]. Therefore, the determination of the amount of unmodified (available) and modified (unavailable) lysine is used as an indicator for the evaluation of food processing on the nutritional quality of food proteins [32, 35, 36].

1.3   The Chemistry of the Maillard Reaction

Last few decades have seen significant progresses in studying the chemistry of the Maillard reaction [3, 4, 26, 32, 37–42]. It is now established that all reactions during the process of Maillard reaction can occur simultaneously and they can be influenced by each other as well as by milieu parameters [6, 27]. In order to better understand the complex processes of the Maillard reaction, Hodge simplifies the process and subdivides the Maillard reaction into three stages chemically:

early stage, intermediate stage, and final stage [3, 26, 43], and the scheme is still widely adopted in contemporary research (Fig. 1.1).

The early stage includes sugar–amine condensation, the formation of a Schiff base and the Amadori rearrangement products (ARPs) [3, 27]. The condensation of the amino compound with the carbonyl group of reducing sugar forms glycosylamine, the most important intermediate compound in Maillard reaction. And the product of the glycosylation subsequently rearranges and dehydrates via deoxyosones, followed by further cyclization to form the important products 1-amino-1-deoxy-2-ketoses and other ARPs [26, 44, 45]. The acid-catalyzed mechanism of the Amadori rearrangement reaction (Fig. 1.2) is supported by the fact that if the hydroxyl group at C-2 is blocked, rearrangement stops. ARPs derived from free amino acids have been detected in various foods such as dried fruits, vegetables, and honey [6, 43, 46]. In the case of protein-containing foods, the ε-amino group of lysine is the primary target of attack by reducing sugars to generate N-ε-ketosyllysine derivatives [47–49].

The intermediate stage of the Maillard reaction is characterized by sugar dehydration, fragmentation, and amino acid degradation. The relatively stable ARPs may dehydrate during severe heating or prolonged storage and be converted

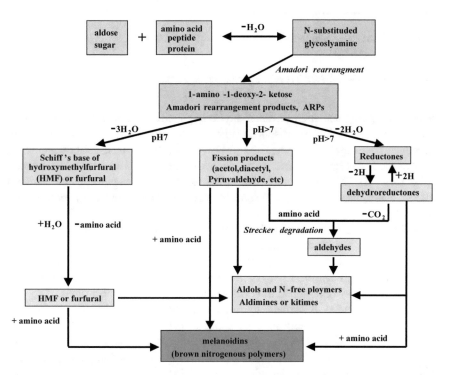

Fig. 1.1 General pathway of Maillard reaction according to Hodge [43] and the MRPs are shown in the frame with different colors corresponding to the products of three stages: early, intermediate, and final stage

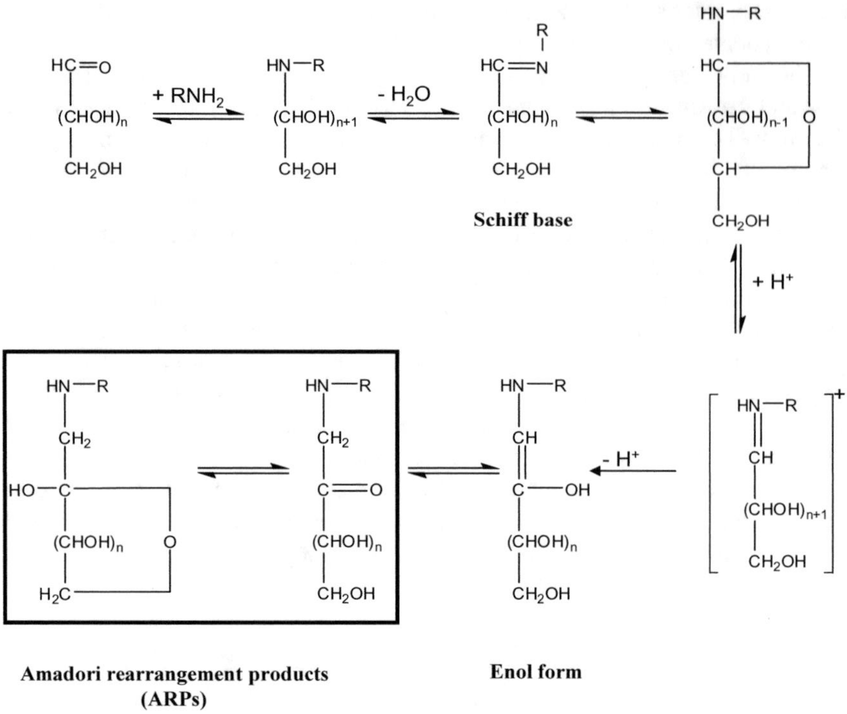

HC=O
(CHOH)$_n$
CH$_2$OH

+ RNH$_2$

HN—R
(CHOH)$_{n+1}$
CH$_2$OH

- H$_2$O

HC=N
|
R
(CHOH)$_n$
CH$_2$OH

Schiff base

HN—R
HC
(CHOH)$_{n-1}$ O
CH
CH$_2$OH

+ H$^+$

HN—R
CH$_2$
HO—C
(CHOH)$_n$ O
H$_2$C

HN—R
CH$_2$
C=O
(CHOH)$_n$
CH$_2$OH

HN—R
CH
C—OH
(CHOH)$_n$
CH$_2$OH

- H$^+$

[HN—R
CH
(CHOH)$_{n+1}$
CH$_2$OH]$^+$

**Amadori rearrangement products
(ARPs)**

Enol form

Fig. 1.2 Mechanism of Amadori rearrangement in the Maillard reaction

upon further fragmentation to dicarbonyl compounds (Fig. 1.3). Dicarbonyl compounds are the major precursors for the formation of various important flavor products, heterocyclic compounds, and polymers [50–52] in the following stage. One of the noted intermediates is the aldehydes formed from Strecker degradation (Fig. 1.4). As Strecker aldehydes can react with sugar-derived Maillard intermediates, they are thought to contribute considerably to the aroma of food [23, 53].

In the final stage, low-molecular weight heterocyclic compounds (Fig. 1.5) and high-molecular weight compounds of polymers form from reactive dicarbonyl and aldehyde intermediates [32, 50, 54]. For example, with the effective catalysis by amines, aldehydes can react with each other by aldol condensation and lead to the formation of many odor-active molecules which contribute to the aromas of cooked food, such as furans, pyrazines, pyrroles, oxazoles, thiazole (when cysteine was involved), and other heterocyclic compounds (Fig. 1.6). Since 1980, advanced glycation end products (AGEs), the stable peptide-bound amino acid derivatives from lysine and arginine side chains of proteins, have attracted increasing attention due to their pathophysiological role in diabetes [55–58].

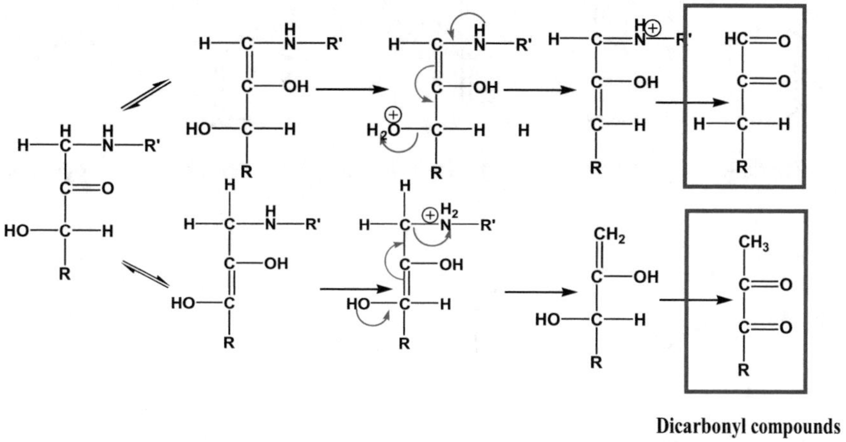

Fig. 1.3 Fragmentation reactions from the ARPs

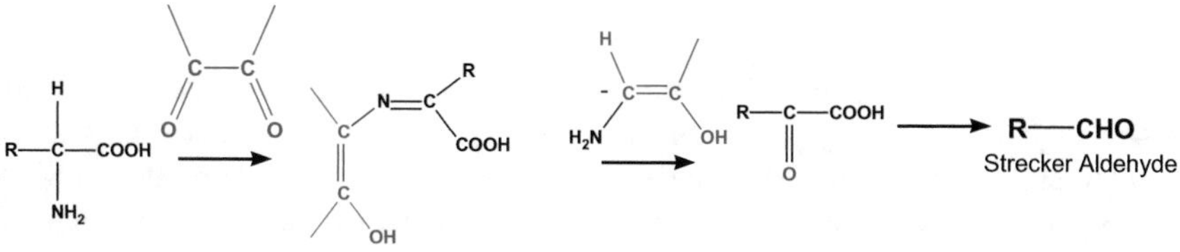

Fig. 1.4 Mechanism of Strecker degradation in the Maillard reaction

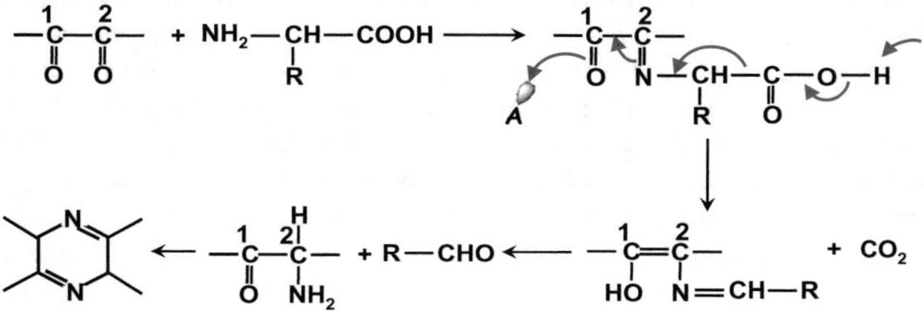

Fig. 1.5 Mechanism of formation of heterocyclic compounds

1.4   Conditions Affecting the Maillard Reaction

The rate of Maillard reactions is heavily affected by the properties of the partici-
pating reactants. For example, pentose sugars (e.g., ribose) react more readily than
hexoses (e.g., glucose) which, in turn, are more reactive than disaccharides
(e.g., lactose). Reaction conditions likewise play an important role in determining

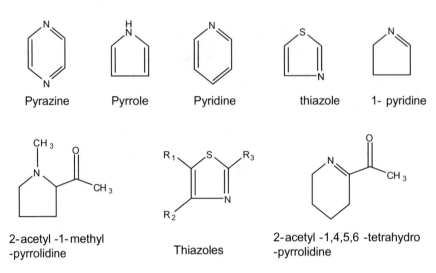

Fig. 1.6 Typical N-heterocyclic compounds produced in the final stage of the Maillard reaction

the rate and outcome of the reaction. The reaction also occurs less readily in foods with a high water activity (a_w) value as water is produced in the chemical process. In addition, the reactants are diluted at high water activity. In contrary, the mobility of reactants is limited at low water activity, despite their presence at increased concentration [59, 60]. In practice, the Maillard reaction occurs most rapidly at intermediate water activity values (0.5–0.8).

The Maillard reaction is also dependent on the pH; in low pH (<6), the formation of furfurals from ARPs is favored, whereas the routes to reductones and fission products are preferred at a high pH (>6) [61]. However, in food preparation and presentation, consumers tend not to accept the addition of chemicals that adjust pH; it is also difficult to control water activity in daily life. For potential applications in the food industry, the present study focuses on the effects of reactants, reaction temperature, reaction time, and heating methods on the Maillard reaction.

Reactions can be induced in either wet- (in solution) or dry-state heating. Wet-heating is a common method used firstly by Louis Maillard [2, 62], whereas dry-heating was initially used in the 1950s [63, 64] and gradually became more popular [46, 65–67]. Dry-heating is accomplished by lyophilizing a solution mixture of protein and reducing sugar, followed by heating under specific temperature and relative humanity. It is a widely accepted and well-developed procedure for the studies of the Maillard reaction [46, 68–72]. In comparison with wet-heated products, dry-heated MRPs exhibit longer shelf life and better functional and thermal properties, whereas protein conformations tend to be less changed than those in solution [73–76]. Dry-heating also requires lower reaction temperature and the extent of Maillard reaction in food modification is more easily controlled; additionally, less time and space is required to obtain the desired results. However, as a drawback, the lyophilization process is not practical in the food industry.

Recently, a new procedure shows that the Maillard reaction can occur readily and efficiently in a protein-reducing sugar system in vacuo [68, 77]. In this study, both heating methods are used to compare the reaction rate and glycoforms in the Maillard reaction to get more understanding about the effect of reaction condition.

Temperature affects markedly the reaction rate and products in the Maillard reaction markedly as have been observed in previous studies [78, 79]. At room temperature or below, the rate of enzymatic reaction is faster than that of non-enzymatic reactions; it might take weeks or even years before significant amounts MRPs can be detected [80]. Maillard reaction through wet-heating therefore needs longer time or higher temperature to be detected as the boiling point of water limits the cooking temperature to 100 °C or below. For example, the surfaces of roasted meats become dehydrated during cooking, allowing Maillard browning to take place while the interior remains moist. The Maillard reaction becomes significant, if the temperature is higher than 100 °C (e.g., during cooking through toasting or frying); however, it is hard to control the rate of the reaction and the reaction products.

1.5   The Analysis of MRPs

A variety of analytical techniques have been adopted to analyze the complex array of Maillard reaction products from simple amino acid–sugars to glycated proteins in foods or biological systems [81–85]. Traditional analytical methods are used to obtain overall information about the bulk changes that occur during Maillard reaction such as fluorescence measurements [86–88], absorbance measurements at 420 nm [60, 89, 90], furosine assay [91–93], and fluorescamine assay [94, 95]. However, these techniques that focus on overall changes cannot provide detailed information on MRPs. Further, ELISA [45, 96–98], NMR [99], affinity chromatographic techniques [100–102], antibody specific chromatography [103–105], and capillary electrophoresis [82, 106] have been used to obtain more detailed data of certain kind of MRPs by specific absorbance or retention times.

Currently, mass spectrometry (MS)-based techniques [107, 108] are employed to characterize the reaction products and examine the structural changes, binding properties, and modification sites of proteins. MS has proven a powerful analytical technique for the determination and characterization of MRPs in foodstuffs and in biological systems [109–111]. Gas chromatography/mass spectrometry (GC-MS) was first used to investigate the Maillard reaction in food products [112–114] and applied to study the formation and decomposition of MRPs during cocoa processing [115, 116]. An isotopic FAB-MS/MS method was later developed to avoid the derivation step and maintain high selectivity in the characterization of deoxyfructosylglycine in model systems [117].

Although quick and accurate, these methods are unable to investigate complex mixtures. To reduce sample complexity, online-coupled capillary electrophoresis (CE) [82] and high-performance liquid chromatography (HPLC) [99, 118] allow the rapid and sensitive separation of samples and improve the limit of detection (LOD) of target analytes in complex mixtures [119]. Explosive growth in the characterization of biological molecules has paralleled the advent of innovative atmospheric pressure ionization techniques, like electrospray ionization (ESI) and matrix-assisted laser desorption/ionization (MALDI) in time-of-flight (TOF) mass spectrometer [85, 120], which are the most widely adopted MS techniques.

1.5.1 *ESI-MS and MALDI-TOF-MS*

ESI-MS and MALDI-TOF-MS techniques are developed by Fenn [121, 122] and Tanaka [123, 124], respectively, in the 1980s. Since then, a large variety of MS instrumentation have become commercially available. The coupling of MS with reverse phase and normal phase HPLC renders it amendable for analyzing a wide range of compounds with different hydrophobicity [125–128]. Hence, MS-based analytical methods are gradually becoming one of the most common methods for identifying and characterizing the MRPs [83, 84, 129, 130].

ESI-MS is a soft desorption technique for biomolecular analysis that had revolutionized the study of polar species in aqueous-based solutions [131] (Fig. 1.7). One of the advantages of ESI lies on the fact that it can reflect the solution chemistry of the analytes.

The solution structure of target compound is preserved in the gas phase, and the relative ion abundances measured in the mass spectrum can reflect their relative concentrations in solution. Another attractive feature of ESI-MS spectra is within a narrow mass range for multiply charged large biomolecules.

MALDI-MS is another soft ionization technique in which the analytes are dried and co-crystallized with an organic matrix compound (a conjugate aromatic structure) on a stainless-steel MALDI plate prior to laser ablation [123, 133]. The analyte-doped matrix crystal evaporates or sublimates via rapid heating on a very

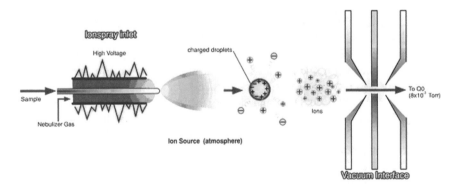

Fig. 1.7 Typical configuration of an ESI source. Adapted with permission from Siu [132]

short timescale to lose its neutral matrix molecules; subsequent proton transfer reactions occur to produce mainly singly charged protonated analytes. The efficiency of absorption depends on the degree of crystallization and the type of matrix used. Nevertheless, the detailed mechanism of ion desorption during MALDI awaits a subject for further investigation [134, 135]. Figure 1.8 presents a schematic representation of ion formation in MALDI.

Compared with ESI-MS, the higher salt and buffer tolerance of MALDI-MS are advantageous for the analysis of biological samples, which usually contains impurities as a result of sample extraction and handling. Singly charged ions are formed predominantly during the MALDI process, thereby allowing simple and accurate molecular weight determination of large compounds.

1.5.2 MS Characterization of ARPs

The early identification of volatile compounds in coffee was initially reported in the 1960s by traditional chemistry methods [112, 114]. From the 1980s onward, research in this field accelerated greatly by the use of GC-MS [12, 136, 137]. Volatile compounds in coffee during roasting were separated by MS-based instruments into approximately 400 MRPs [116, 138–141]. Nowadays, MS- and MS-related techniques are widely used to identify MRPs in all kinds of food [142]. Recently, LC-MS has been used to identify MRPs and even flavor compounds formed in model Maillard reactions [143]. MS/MS-based techniques pursue structure identification not only on the basis of the molecular ion but also on all fragments. The valuable information of fragmentation is directly linked to the

Fig. 1.8 Ion formation in MALDI

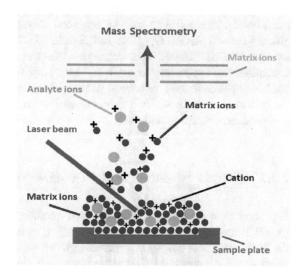

chemical structure, thus allowing a higher degree of confidence in the identification of unknown volatile and nonvolatile compounds.

ARPs produce reactive intermediate compounds under physiological conditions and during food processing and storage which play a critical role in the formation of aroma, taste, and color at later reaction stages. Likewise, some compounds produced from ARPs decrease the nutritive value and form toxic compounds in later stages. These compounds have attracted particular attention due to the marked effect on human health [144, 145].

With the development of MS techniques, the formation of ARPs and the reaction rate in the process of Maillard reaction can be monitored and characterized by both MALDI-MS and ESI-MS. MS has been used extensively in the qualitative and quantitative evaluation of the distribution and loading values of small molecules, such as sugars, drugs, and pesticides, that are covalently bound to amino acids, peptides, and proteins [146]. MS has been successfully applied to study both qualitative and semiquantitative analyses of MRPs and opens up research possibilities of understanding the formation and degradation of ARPs during the early stages of Maillard reaction. For example, the number of hexose residues coupled to a protein or peptide is estimated by a mass increase of 162 Da per attached monosaccharide. In addition to determining their molecular weights, structural elucidation of these compounds is also possible through tandem mass spectrometry (MS/MS).

Various approaches have been evaluated for the identification of ARPs, such as GC-MS with volatile sample derivatization prior to analysis and HPLC with nonvolatile compounds without prior derivatization [147–149]. However, the analysis of ARPs is still a challenging task, because ARPs are numerous and their structures are similar. Recently, Davidek et al. [150] report a method based on high-performance anion exchange chromatography coupled with an electrochemical and/or diode array detector to detect and monitor known ARPs. However, unambiguous identification of ARPs can hardly be achieved with this method. Fast atom bombardment (FAB)-MS/MS has been successfully used to identify a few glucose-derived ARPs. Over the last decade, FAB has been almost completely replaced by ESI-MS/MS. This technique offers the particular advantages of a soft ionization, leading mainly to $[M + H]^+$ ions with very limited fragmentation and has been applied to analyze several Amadori compounds [78, 151–153]. However, the systematical study of ARPs in amino acid–sugar and peptide–sugar systems for structural determination and study of fragmentation behavior is still required.

1.5.3   Protein Glycation and Glycation Site Analysis

Glycation is a common posttranslational modification found in food proteins [6, 42]. It readily occurs in food proteins during storage [154] and cooking through the Maillard reaction [60, 92]. Protein glycation is achieved by early Maillard reaction (condensation of reducing sugars with amino acids amino groups, i.e., mainly lysine residues), and the effect of glycation on food proteins may be beneficial or detrimental, depending on the extent of glycation. Controlled, limited glycation has been reported to impart beneficial effects to proteins, whereas extensive glycation results in protein cross-linking and loss of protein solubility and nutritional value [155, 156]. Assessment of protein glycation extent in biological or food systems can be performed using various analytical methods such as furosine assay (formed upon acid hydrolysis of glycated lysine residues), free amino groups (lysine and N-terminal residues) assessment, or fluorescamine assay [94, 155].

MS is a common tool used to evaluate and characterize protein glycation, both in food and in biological systems. ESI-MS and MALDI-TOF-MS provide accurate relative molecular mass values for the analysis of protein glycation through the specific mass shift they impart on the residue. For example, the formation of ARPs occurred during protein glycation can be monitored by the detection of a mass shift of 162 Da. Similarly, MS has been used to study lactosylation of α-lactoglobulin in milk during processing [157, 158] and to display the association behavior of glycated proteins [159, 160]. The detection of protein glycation products and further identification of glycation sites is a challenging area of investigation, one of the key reasons being that glycated peptides and proteins decrease the capacity of ionization compared with the corresponding unmodified peptides, and the conventional proteolytic method of mass mapping is somewhat limited in the modification of glycation [161]. ESI-MS and MALDI-TOF-MS have been extensively used to characterize posttranslational modifications of proteins in the Maillard reaction [1, 40, 49, 85, 111, 162–165]. Lysozyme, a well-known food protein, has seven free amino groups (one α-NH$_2$ at the N-terminal and six ε-NH$_2$ of lysine residue), two of them have been shown to be available to participate in the Maillard reaction [166, 167] and are identified as the N-terminal α-NH$_2$ at Lys1 and ε-NH$_2$ at Lys97 [168]. Three amino groups in lysozyme modified with dextran have been identified by peptide mapping, molecular modeling, and trypsin digestion, and the sites were the α-NH$_2$ of Lys1 and two ε-NH$_2$ of Lys33 and Lys97 [73, 169].

HPLC and CE separation techniques coupled with MS have enabled the analysis of various ARPs in model Maillard reaction mixtures [111, 120]. For example, CE-ESI-MS has been used to determine MRPs in skim milk powders [170]. LC-ESI-MS has been applied to characterize the ARPs obtained after enzymatic hydrolysis of whey proteins and to monitor the nutritional loss caused by the blockage of lysine residues during milk processing [149, 171]. MALDI-TOF-MS had also been used to analyze ARPs in protein–sugar systems [172]. Similar methods have been developed for analysis the glycation-related diseases such as diabetes, and to study other types of PTMs, such as oxidation and methylation [40, 173, 174].

More recently, methods were developed based on the mass-to-charge ratio of precursor ion scanning, and particular target ions were identified for detection and characterization by LC-MS/MS [175, 176]. Meanwhile, high-resolution MALDI-TOF/TOF-MS has been used to obtain structural data of glycation sites from the fragmentation ions [177–179].

1.6   Summary

The research is to study the effects of reaction conditions on the Maillard reaction, to characterize the structure of specific ARPs by NMR spectroscopy, to investigate fragmentation behavior of ARPs, to monitor the Maillard reaction by ESI-MS and MALDI-TOF-MS, and to develop a method to characterize the modification sites of protein using ESI and MALDI tandem mass spectrometry. In addition, the effects of glucosylation level on protein functional and thermal properties and the structural and glycoforms changes in the modified protein are also examined.

In the present study, the Maillard reactions of lysozyme–glucose system are monitored during both wet- and dry-heating under controlled temperatures. MALDI-TOF-MS and ESI-MS are used to monitor the Maillard reaction and display the extent of glycation in lysozyme. The glycated lysozyme is then purified and digested, after which MALDI-TOF/TOF-MS/MS and nano-LC-QqTOF-MS/ MS are used to identify all glycation sites of lysozyme. Three temperatures (50, 70, and 90 °C) are selected for both wet-heating and dry-heating in amino acid–sugar, peptide–sugar, and protein–sugar systems.

The fragmentation behavior of well-defined amino acid and peptide-related ARPs based on the structure identification by NMR spectroscopy and theoretical calculation is studied. Furthermore, the influence of structures including the sugar skeleton and position of glycation on fragmentation behavior, and the correlation between fragmentation and their structures based on MS/MS are investigated. The main fragmentation pathways of sugar and peptide moieties are summarized in the present study. Moreover, characteristic fragment ions are successfully employed to sequence ARPs of peptides and identify the glycation sites of peptides and proteins.

The effects of glycation on the emulsifying and thermal properties of lysozyme are evaluated. A methodology is developed which combines different mass spectrometric techniques to monitor modifications in early Maillard reaction. In addition to examining the effects of different number of attached sugar moieties on protein properties, the unequivocal location of Maillard modification sites on the protein backbone is also demonstrated. A systematic approach is adopted to study the effect of amino acids and lysine-containing peptides on the rate of the Maillard reaction. Based on the mechanism of the Maillard reaction, six prototypical amino acids (lysine, arginine, asparagine, glutamine, histidine, and tryptophan) and their N-terminal acetylated forms are selected in amino acid–sugar systems. These amino acids all contain either an amino or imino group on their side chains except glycine is used as a control. In peptide–sugar systems, factors affecting the Maillard reaction rate including the length of peptides and position of lysine are studied.

The formation of key ARPs is systematically studied beginning with simple amino acid–sugar model and peptide–sugar model, and proceeding to protein–sugar system. MS–based techniques are combined with NMR spectroscopy and theoretical calculation to obtain unequivocal structure and fragmentation behaviors of the ARPs, which serves to monitor the ARPs in complex systems. The glycoforms and glycation sites are analyzed. The studies on extent and sites of glycation in protein are also helpful for understanding the changes in thermal and functional properties. The results obtained in the present study could lead to fundamental understanding of the Maillard reaction and gas phase chemistry, leading to an improvement of the current protein identification methods. A better understanding of the reactants and reaction condition will enable design and production of food modifications under optimal conditions. The combination of MS and NMR data provides an attractive option for studying ARPs. Specific fragmentation ions allow facile characterization of ARPs with furthered understanding of ARP fragmentation behaviors in MS. An easily adoptable methodology to monitor Maillard reactions and identify modification sites in food proteins will expand the use of MS-based techniques in the food industries and provide better understanding of the structural and conformational changes in the modified proteins. Finally, a direct method to modify protein with sugars can lead to enhanced functional properties for food application.

References

1. Maillard LC (1912) Reaction of amino acids on sugars: formation of melanoidins by a systematic way. Compt Rend Acad Sci 154:66–68
2. Maillard LC (1912) General reaction between amino acids and sugars: the biological consequences. Compt Rend Soc Biol 72:599–601
3. Hodge JE (1955) The amadori rearrangement. Adv Carbohydr Chem 10:169–205
4. Billaud C, Adrian J (2003) Louis-Camille Maillard, 1878–1936. Food Rev Inter 19(4): 345–374
5. Dworschak E, Carpenter KJ (1980) Non-enzyme browning and its effect on protein nutrition. Crc Rev Food Sci Nutr 13(1):1–40
6. Friedman M (1996) Food browning and its prevention: an overview. J Agric Food Chemy 44 (3):631–653
7. Lee KW, Mossine V, Ortwerth BJ (1998) The relative ability of glucose and ascorbate to glycate and crosslink lens proteins in vitro. Experi Eye Res 67(1):95–104
8. Friedman M, Brandon DL (2001) Nutritional and health benefits of soy proteins. J Agric Food Chem 49(3):1069–1086
9. Talasz H, Wasserer S, Puschendorf B (2002) Nonenzymatic glycation of histones in vitro and in vivo. J Cellu Biochem 85(1):24–34
10. Ledl F, Schleicher E (1990) New aspects of the maillard reaction in foods and in the human-body. Angew Chem Inter 29(6):565–594
11. Mottram DS (1994) Flavor compounds formed during the maillard reaction. Therm Gener Flavors 543:104–126
12. Vernin G, Metzger J, Obretenov T et al (1988) GC/MS (EI, PCI, SIM) data bank analysis of volatile compounds arising from thermal degradation of glucose-valine amadori intermediates. Elsevier Scie, Amsterdam
13. Ghiggeri GM, Candiano G, Ginevri F et al (1988) Spectrophotometric determination of browning products of glycation of protein amino-groups based on their reactivity with nitro blue tetrazolium salts. Analyst 113(7):1101–1104
14. Morgan F, Molle D, Henry G et al (1999) Glycation of bovine beta-Lactoglobulin: effect on the protein structure. Int J Food Sci Tech 34(5–6):429–435
15. Miyaguchi Y, Tsutsumi M, Nagayama K (1999) Properties of glycated globin prepared through the Maillard reaction. J Japan Soc Food SciTech 46(8):514–520

16. Lopez-Fandino R (2006) Functional improvement of milk whey proteins induced by high hydrostatic pressure. Critical Rev Food Sci Nutri 46(4):351–363

17. Nakamura S, Suzuki Y, Ishikawa E et al (2008) Reduction of in vitro allergenicity of buckwheat Fag e 1 through the Maillard-type glycosylation with polysaccharides. Food Chem 109(3):538–545

18. Swaisgood HE, Catignani GL (1982) In vitro measurement of effects of processing on protein nutritional quality. J Food Prot 45(13):1248–1256

19. Friedman M (1999) Chemistry, biochemistry, nutrition, and microbiology of lysinoalanine, lanthionine, and histidinoalanine in food and other proteins. J Agric Food Chem 47(4): 1295–1319

20. Vanderhaegen B, Neven H, Verachtert H et al (2006) The chemistry of beer aging—a critical review. Food Chem 95(3):357–381

21. Guenther H, Anklam E, Wenzl T et al (2007) Acrylamide in coffee: review of progress in analysis, formation and level reduction. Food Add Contam 24:60–70

22. Frye EB, Degenhardt TP, Thorpe SR et al (1998) Role of the Maillard reaction in aging of tissue proteins—advanced glycation end product-dependent increase in imidazolium cross-links in human lens proteins. J Biol Chem 273(30):18714–18719

23. Yaylayan VA (2003) Recent advances in the chemistry of Strecker degradation and Amadori rearrangement: implications to aroma and color formation. Food Sci Tech Res 9(1):1–6

24. Gandemer G (1999) Lipids and meat quality: lipolysis, oxidation, Maillard reaction and flavour. Sci Des Aliments 19(3–4):439–458

25. Slaughter JC (1999) The naturally occurring furanones: formation and function from pheromone to food. Biol Rev 74(3):259–276

26. Hodge JE, Rist CE (1953) The Amadori rearrangement under new conditions and its significance for non-enzymatic browning reactions. J Am Chem Soc 75(2):316–322

27. Wrodnigg TM, Eder B (2001) The Amadori and Heyns rearrangements: landmarks in the history of carbohydrate chemistry or unrecognized synthetic opportunities? Glyoscience 215:115–152

28. Adrian J, Godon B, Petit L (1962) Nutrition—Sur De Nouvelles Incidences Nutritionnelles De La Reaction De Maillard. Compt Rend Acad Sci 255(2):391

29. Finot PA (2005) Historical perspective of the Maillard reaction in food science. Maillard React 1043:1–8

30. Somoza V (2005) Five years of research on health risks and benefits of Maillard reaction products: an update. Mol Nutr Food Res 49(7):663–672

31. Meade SJ, Reid EA, Gerrard JA (2005) The impact of processing on the nutritional quality of food proteins. J AOAC Inter 88(3):904–922

32. Ferrer E, Alegria A, Farre R et al (1999) Review: indicators of damage of protein quality and nutritional value of milk. Food Sci Tech Inter 5(6):447–461

33. Birlouez-Aragon I, Charissou A, Damjanovic S (2006) Milk fortification and processing—nutritional impact. Sci Des Aliments 26(6):483–491

34. Meltretter J, Pischetsrieder M (2008) Application of mass spectrometry for the detection of glycation and oxidation products in milk proteins. Rec Adv Food and Biomed Sci 1126:134–140

35. Elango R, Humayun MA, Ball RO et al (2007) Lysine requirement of healthy school-age children determined by the indicator amino acid oxidation method. Am J Clin Nutr 86: 360–365

36. Elango R, Ball RO, Pencharz PB (2008) Individual amino acid requirements in humans: an update. Clin Nutr Meta Care 11:34–39

37. Hodge JE, Rist CE (1952) N-glycosyl derivatives of secondary amines. J Am Chem Soc 74 (6):1494–1497

38. Tran QD, Hendriks WH, van der Poell AFB (2008) Effects of extrusion processing on nutrients in dry pet food. J Sci Food Agric 88(9):1487–1493

39. Zhang Y (2007) Formation and reduction of acrylamide in Maillard reaction: a review based on the current state of knowledge. Cri Rev Food Sci Nutr 47(5):521–542

40. Lapolla A, Fedele D, Seraglia R et al (2006) The role of mass spectrometry in the study of non-enzymatic protein glycation in diabetes: an update. Mass Spectrom Rev 25(5):775–797

41. Abul Farah M, Bose S, Lee JH et al (2005) Analysis of glycated insulin by MALDI-TOF mass spectrometry. Biochim Biophys Acta 1725(3):269–282

42. Singh R, Barden A, Mori T et al (2002) Advanced glycation end-products: a review. Diabetologia 45(2):293

43. Hodge JE (1953) Dehydrated foods—chemistry of browning reactions in model systems. J Agric Food Chem 1(15):928–943

44. Tjan SB, Ouweland GA (1974) Pmr investigation into structure of some N-substituted 1-amino-1-deoxy-d-fructoses (amadori rearrangement products)—evidence for a preferential conformation in solution. Tetrahedron 30(16):2891–2897

45. Kislinger T, Humeny A, Peich CC et al (2003) Relative quantification of N-epsilon-(carboxymethyl)lysine, imidazolone A, and the Amadori product in glycated lysozyme by MALDI-TOF mass spectrometry. J Agric Food Chem 51(1):51–57

46. Kato A, Mifuru R, Matsudomi N et al (1992) Functional casein-polysaccharide conjugates prepared by controlled dry heating. Biosci Biotech Biochem 56(4):567–571

47. Ahmed MU, Thorpe SR, Baynes JW (1986) Identification of N-epsilon-carboxymethyllysine as a degradation product of fructoselysine in glycated protein. J Biol Chem 261(11):4889–4894

48. Biemel KM, Lederer MO (2003) Site-specific quantitative evaluation of the protein glycation product N-6-(2,3-dihydroxy-5,6-dioxohexyl)-L-lysinate by LC-(ESI)MS peptide mapping: evidence for its key role in AGE formation. Bioconj Chem 14(3):619–628

49. Saraiva MA, Borges CM, Florencio MH (2006) Reactions of a modified lysine with aldehydic and diketonic dicarbonyl compounds: an electrospray mass spectrometry structure/activity study. J Mass Spectrom 41(2):216–228

50. Obretenov T, Vernin G (1998) Melanoidins in the maillard reaction. Food Flav 40:455–482

51. Riha WE, Ho CT (1998) Flavor generation during extrusion cooking. Food Chem 434:297–306

52. Martins S, Jongen WMF, van Boekel M (2000) A review of Maillard reaction in food and implications to kinetic modelling. Tren Food Sci Tech 11(9–10):364–373

53. Cheng KW, Chen F, Wang MF (2006) Heterocyclic amines: chemistry and health. Mol Nutr Food Res 50(12):1150–1170

54. Faist V, Erbersdobler HF (2001) Metabolic transit and in vivo effects of melanoidins and precursor compounds deriving from the Maillard reaction. Ann Nutr Meta 45(1):1–12

55. Radoff S, Vlassara H, Cerami A (1987) Isolation and characterization of a receptor for proteins modified by advanced glycation end-products (age) from a macrophage cell-line. Age 10(4):163

56. Yamagishi SI, Ueda S, Okuda S (2007) Food-derived advanced glycation end products (AGEs): a novel therapeutic target for various disorders. Curr Pharm Design 13:2832–2836

57. Thornalley PJ (1998) Cell activation by glycated proteins. AGE receptors, receptor recognition factors and functional classification of AGEs. Cel Molr Biol 44(7):1013–1023

58. Sato T, Iwaki M, Shimogaito N et al (2006) TAGE (toxic AGEs) theory in diabetic complications. Cur Mol Med 6(3):351–358

59. Chen WS, Liu DC, Chen MT (2002) Effects of high level of sucrose on the moisture content, water activity, protein denaturation and sensory properties in Chinese-style pork jerky. Asian-Aus J Anim Sci 15(4):585–590

60. Davies CGA, Netto FM, Glassenap N et al (1998) Indication of the Maillard reaction during storage of protein isolates. J Agric Food Chem 46(7):2485–2489

61. Davidek T, Clety N, Aubin S et al (2002) Degradation of the Amadori compound N-(1-deoxy-D-fructos-1-yl)glycine in aqueous model systems. J Agric and Food Chem 50 (19):5472–5479

62. Maillard LC (1916) A synthesis of humic matter by effect of amine acids on sugar reducing agents. Ann Chim Fra 5:258–316

63. Lea CH, Hannan RS (1950) Studies of the reaction between proteins and reducing sugars in the dry state 3. Nature of the protein groups reacting. Biochim Biophys Acta 5(3):433–454
64. Lea CH, Hannan RS (1950) Studies of the reaction between proteins and reducing sugars in the dry state 2. Further observations on the formation of the casein-glucose complex. Biochim Biophys Acta 4(4):518–531
65. Kato A, Ibrahim HR, Watanabe H et al (1989) New approach to improve the gelling and surface functional-properties of dried egg-white by heating in dry state. J Agric Food Chem 37(2):433–437
66. Kato A, Sasaki Y, Furuta R et al (1990) Functional protein polysaccharide conjugate prepared by controlled dry-heating of ovalbumin dextran mixtures. Agric Biol Chem 54 (1):107–112
67. Kato Y, Matsuda T, Nakamura R (1993) Improvement of physicochemical and enzymatic-properties of bovine trypsin by nonenzymatic glycation. Biosci Biotech Biochem 57(1):1–5
68. Kaplan H, Taralp A (1997) Nonaqueous chemical modification of lyophilized proteins. Tech Protein Chem 8:219–230
69. Hattori M (2002) Functional improvements in food proteins in multiple aspects by conjugation with saccharides: case studies of beta-lactoglobulin-acidic polysaccharides conjugates. Food Sci Tech Res 8(4):291–299
70. Pham VT, Ewing E, Kaplan H et al (2008) Glycation improves the thermostability of trypsin and chymotrypsin. Biotech Bioengineer 101(3):452–459
71. Achouri A, Boye JI, Yaylayan VA et al (2005) Functional properties of glycated soy 11S glycinin. J Food Sci 70(4):C269–C274
72. Sola RJ, Griebenow K (2006) Chemical glycosylation: new insights on the interrelation between protein structural mobility, thermodynamic stability, and catalysis. FEBS Lett 580 (6):1685–1690
73. Yeboah FK, Alli I, Yaylayan VA et al (2004) Effect of limited solid-state glycation on the conformation of lysozyme by ESI-MSMS peptide mapping and molecular modeling. Bioconj Chem 15(1):27–34
74. Van Der Veen M, Norde W, Stuart MC (2005) Effects of succinylation on the structure and thermostability of lysozyme. J Agric Food Chem 53(14):5702–5707
75. Aoki T, Kitahata K, Fukumoto T et al (1997) Improvement of functional properties of beta-lactoglobulin by conjugation with glucose-6-phosphate through the Maillard reaction. Food Res Inter 30(6):401–406
76. Broersen K, Voragen AGJ, Hamer RJ et al (2004) Glycoforms of beta-lactoglobulin with improved thermostability and preserved structural packing. Biotech Bioengineer 86(1): 78–87
77. Kato A (2002) Industrial applications of Maillard-type protein-polysaccharide conjugates. Food Sci Tech Res 8(3):193–199
78. Mennella C, Visciano M, Napolitano A et al (2006) Glycation of lysine-containing dipeptides. J Pep Sci 12(4):291–296
79. Machiels D, Istasse L (2002) Maillard reaction: importance and applications in food chemistry. Ann Med Veter 146(6):347–352
80. O'Brien J, Nursten MJ, Ames M (1988) The Maillard reaction in foods and medicine
81. Chen TM, Coutant JE (1989) Thermospray high performance liquid chromatographic mass spectrometric characterization of biological macromolecules. J Chromatogr A 463(1): 95–106
82. Frazier RA, Ames JM, Nursten HE (1999) The development and application of capillary electrophoresis methods for food analysis. Electrophoresis 20(15–16):3156–3180
83. Yeboah FK, Yaylayan VA (2001) Analysis of glycated proteins by mass spectrometric techniques: qualitative and quantitative aspects. Nahrung-Food 45(3):164–171
84. Zhang Y, Zhang GY (2005) Occurrence and analytical methods of acrylamide in heat-treated foods—review and recent developments. J Chromatogr A 1075(1–2):1–21

85. Yeboah FK, Alli I, Yaylayan VA et al (2000) Monitoring glycation of lysozyme by electrospray ionization mass spectrometry. J Agric Food Chem 48(7):2766–2774

86. Karoui R, Bosset JO, Mazerolles G et al (2005) Monitoring the geographic origin of both experimental French Jura hard cheeses and Swiss Gruyere and L'Etivaz PDO cheeses using mid-infrared and fluorescence spectroscopies: a preliminary investigation. Inter Dairy J 15 (3):275–286

87. Karoui R, Dufour E, Pillonel L et al (2005) The potential of combined infrared and fluorescence spectroscopies as a method of determination of the geographic origin of Emmental cheeses. Inter Dairy J 15(3):287–298

88. Andersen CM, Mortensen G (2008) Fluorescence spectroscopy: a rapid tool for analyzing dairy products. J Agric Food Chem 56:720–729

89. Saltmarch M, Vagniniferrari M, Labuza TP (1981) Theoretical basis and application of kinetics to browning in spray-dried whey food systems. Prog Food and Nutr Sci 5(1–6): 331–344

90. Kim MN, Saltmarch M, Labuza TP (1981) Non-enzymatic browning of hygroscopic whey powders in open versus sealed pouches. J Food Proc Pres 5(1):49–57

91. Fenaille F, Campos-Gimenez E, Guy PA et al (2003) Monitoring of beta-lactoglobulin dry-state glycation using various analytical techniques. Anal Biochem 320(1):144–148

92. Ramirez-Jimenez A, Guerra-Hernandez EJ, Garcia-Villanova B (2004) Evaluation of amino group losses for monitoring storage of infant cereals containing milk. Food Control 15 (5):351–354

93. Bourais I, Amine A, Moscone D et al (2006) Investigation of glycated protein assay for assessing heat treatment effect in food samples and protein-sugar models. Food Chem 96 (3):485–490

94. Yaylayan VA, Huyghuesdespointes A, Polydorides A (1992) A fluorescamine-based assay for the degree of glycation in bovine serum-albumin. Food Res Inter 25(4):269–275

95. Birlouez-Aragon I, Leclere J, Quedraogo CL et al (2001) The FAST method, a rapid approach of the nutritional quality of heat-treated foods. Nahrung-Food 45(3):201–205

96. Koschinsky T, He CJ, Mitsuhashi T et al (1997) Orally absorbed reactive glycation products (glycotoxins): an environmental risk factor in diabetic nephropathy. Proc Acad Sci Am 94 (12):6474–6479

97. Facchiano F, D'Arcangelo D, Russo K et al (2006) Glycated fibroblast growth factor-2 is quickly produced in vitro upon low-millimolar glucose treatment and detected in vivo in diabetic mice. Mol Endoc 20(11):2806–2818

98. Buongiorno AM, Morelli S, Sagratella E et al (2008) Immunogenicity of advanced glycation end products in diabetic patients and in nephropathic non-diabetic patients on hemodialysis or after renal transplantation. J Endoc Invest 31(6):558–562

99. Olano A, Santamaria G, Corzo N et al (1992) Determination of free carbohydrates and Amadori compounds formed at the early stages of nonenzymatic Browning. Food Chem 43 (5):351–358

100. Saito A, Nagai R, Tanuma A et al (2003) Role of megalin in endocytosis of advanced glycation end products: implications for a novel protein binding to both megalin and advanced glycation end products. J Am Soc Nephr 14(5):1123–1131

101. Lopes-Virella MF, Thorpe SR, Derrick MB et al (2005) The immunogenicity of modified lipoproteins. Maillard React 1043:367–378

102. Zhang QB, Tang N, Brock JWC et al (2007) Enrichment and analysis of nonenzymatically glycated peptides: boronate affinity chromatography coupled with electron-transfer dissociation mass spectrometry. J Proteo Res 6(6):2323–2330

103. Ikeda K, Nagai R, Sakamoto T et al (1998) Immunochemical approaches to AGE-structures: characterization of anti-AGE antibodies. J Immunl Meth 215(1–2):95–104

104. Iwamoto H, Motomiya Y, Miura K et al (2001) Immunochemical assay of hemoglobin with N-epsilon-(carboxymethyl)lysine at lysine 66 of the beta chain. Clin Chem 47(7):1249–1255

105. Nagai R, Fujiwara Y, Mera K et al (2008) Usefulness of antibodies for evaluating the biological significance of AGEs. Maillard React 1126:38–41

106. Steffan W, Balzer HH, Lippert F et al (2006) Characterization of casein lactosylation by capillary electrophoresis and mass spectrometry. Eur Food Res Tech 222(3–4):467–471
107. Feng X, Siegel MM (2007) FTICR-MS applications for the structure determination of natural products. Anal Bioanal Chem 389:1341–1363
108. Murakami T, Fukutsu N, Kondo J et al (2008) Application of liquid chromatography-two-dimensional nuclear magnetic resonance spectroscopy using pre-concentration column trapping and liquid chromatography-mass spectrometry for the identification of degradation products in stressed commercial amlodipine maleate tablets. J Chromatogr A 1181(1–2): 67–76
109. Sabbadin S, Seraglia R, Allegri G et al (1999) Matrix-assisted laser desorption/ionization mass spectrometry in evaluation of protein profiles of infant formulae. Rapid Comm Mass Spectrom 13(14):1438–1443
110. Nibbering NMM (2006) Four decades of joy in mass spectrometry. Mass Spectrom Rev 25 (6):962–1017
111. Tagami U, Akashi S, Mizukoshi T et al (2000) Structural studies of the Maillard reaction products of a protein using ion trap mass spectrometry. J Mass Spectrom 35(2):131–138
112. Viani R, Mugglerc F, Reymond D et al (1965) Sur La Composition De Larome De Cafe. Helvet Chim Acta 48(7):1809
113. Finot PA, Bricout J, Viani R et al (1968) Identification of a new lysine derivative obtained upon acid hydrolysis of heated milk. Experientia 24(11):1097
114. Zlatkis A, Sivetz M (1960) Analysis of coffee volatiles by gas chromatography. J Food Sci 25:395–398
115. Ciner-Doruk M, Eichner K (1979) Formation and stability of amadori compounds in low moisture foods. Z Lebensmittel-Untersuchungund-Forschung 168:9–21
116. Gautschi F, Winter M, Flament I et al (1967) New developments in coffee aroma research. J Agric Food Chem 15:15–23
117. Staempfli AA, Blank I, Fumeaux R et al (1994) Study of the decomposition of the Amadori compound N-(1-deoxy-D-fructos-1-yl)-glycine (DFG) in model systems: quantification by FAB/MS/MS. Biol Mass Spectrom 23:642–646
118. Martins-Junior HA, Wang AY, Alabourda J et al (2008) A validated method to quantify folic acid in wheat flour samples using liquid chromatography—tandem mass spectrometry. J Braz Chem Soc 19(5):971–977
119. Mirsaleh-Kohan N, Robertson WD, Compton RN (2008) Electron ionization time-of-flight mass spectrometry: historical review and current applications. Mass Spectrom Rev 27 (3):237–285
120. Odani H, Matsumoto Y, Shinzato T et al (1999) Mass spectrometric study on the protein chemical modification of uremic patients in advanced Maillard reaction. J Chromatogr B 731 (1):131–140
121. Yamashita M, Fenn JB (1984) Electrospray ion-source—another variation on the free-jet theme. J Phys Chem 88(20):4451–4459
122. Fenn JB, Mann M, Meng CK et al (1989) Electrospray ionization for mass-spectrometry of large biomolecules. Science 246(4926):64–71
123. Tanaka T, Homma Y, Kurosawa S (1988) Secondary ion mass-spectrometric ion yields and detection limits of impurities in indium-phosphide. Anal Chem 60(1):58–61
124. Koichi T, Yutaka I, Satoshi A et al (1988) Protein and polymer analyses up to m/z 100 000 by laser ionization time-of-flight mass spectrometry. Rapid Comm Mass Spectrom 2(8): 151–153
125. Terada H, Tamura Y (2003) Determination of acrylamide in processed foods by column-switching HPLC with UV detection. J Food Hyg Soc Jap 44(6):303–309
126. zur Stadt U, Eckert C, Rischewski J et al (2003) Identification and characterisation of clonal incomplete T-cell-receptor V delta 2-D delta 3/D delta 2-D delta rearrangements by denaturing high-performance liquid chromatography and subsequent fragment collection: implications for minimal residual disease monitoring in childhood acute lymphoblastic leukemia. J Chromatogr B 792(2):287–298

127. Fan Y, Zhang M, Da SL et al (2005) Determination of endocrine disruptors in environmental waters using poly(acrylamide-vinylpyridine) monolithic capillary for in-tube solid-phase microextraction coupled to high-performance liquid chromatography with fluorescence detection. Analyst 130(7):1065–1069

128. Wang HY, Lee AWM, Shuang SM et al (2008) SPE/HPLC/UV studies on acrylamide in deep-fried flour-based indigenous Chinese foods. Microchem J 89(2):90–97

129. Crews C, Castle L (2007) A review of the occurrence, formation and analysis of furan in heat-processed foods. Trends in Food Sci Tech 18(7):365–372

130. Fay LB, Brevard H (2005) Contribution of mass spectrometry to the study of the Maillard reaction in food. Mass Spectrom Rev 24(4):487–507

131. Fenn JB (2003) Electrospray wings for molecular elephants (Nobel lecture). Angew Chem Inter 42(33):3871–3894

132. Siu SO (2008) Biological mass spectrometry of peptides and glycopeptides. Thesis of the degree of Doctor of Philosophy, pp. 88–114

133. Tanaka K, Waki H, Ido Y et al (1988) Protein and polymer analyses up to m/z 100 000 by laser ionization time-of-flight mass spectrometry. Rapid Commun Mass Spectrom 2(8): 151–153

134. Karas M, Gluckmann M, Schafer J (2000) Ionization in matrix-assisted laser desorption/ ionization: singly charged molecular ions are the lucky survivors. J Mass Spectrom 35(1): 1–12

135. Gabelica V, Schulz E, Karas M (2004) Internal energy build-up in matrix-assisted laser desorption/ionization. J Mass Spectrometry 39(6):579–593

136. Petitjean M (1981) Banques de spectres de masse des composes volatils presents dans les aromes alimentaires. Ind Alim Agric 09:741–751

137. Vernin G, Metzger J, Boniface C et al (1992) Kinetics and thermal-degradation of the fructose methionine amadori intermediates—GC-MS/specma data-bank identification of volatile aroma compounds. Carbohydr Res 230(1):15–29

138. Baltes W, Bochmann G (1987) Model reactions on roast aroma formation: reaction of serine and threonine with sucrose under the conditions of coffee roasting and identification of new coffee aroma compounds. J Agric Food Chem 35(3):340–346

139. Baltes W, Bochmann G (1987) Model reactions on roast aroma formation: mass-spectrometric identification of pyrroles from the reaction of serine and threonine with sucrose under the conditions of coffee roasting. Z Lebensm-Unters-Forschung 184(6): 478–484

140. Flament I, Bessie R, Thomas Y (2002) Coffee flavor chemistry. Wiley, Chichester

141. Stoll M, Winter M, Gautschi F et al (1967) Recherches surles aromes. Sur l'arome de cafe I. Helv Chim Acta 50:628–694

142. Newton T, Smith S, Williams T et al (2001) The use of automated accurate mass fragment assignment for the confirmation of library search hits in the GC-MS analysis of complex mixtures. T Chemie Plus 12:24–28

143. Fay LB, Newton A, Simian H et al (2003) Potential of gas chromatography-orthogonal accelerationtime-of-flight mass spectrometry (GC-oaTOFMS) in flavour research. J Agric Food Chem 51:2708–2713

144. Adrian J, Frangne R (1973) Maillard reaction 8. Role of premelanoidins on nitrogen digestibility in-vivo and on proteolyse in-vitro. Ann Nutr Alimen 27(3):111–123

145. Corzo-Martinez M, Moreno FJ, Olano A et al (2008) Structural characterization of bovine beta-lactoglobulin-galactose/tagatose Maillard complexes by electrophoretic, chromatographic, and spectroscopic methods. J Agric Food Chem 56(11):4244–4252

146. Ericson C, Phung QT, Horn DM et al (2003) An automated noncontact deposition interface for liquid chromatography matrix-assisted laser desorption/ionization mass spectrometry. Anal Chem 75:2309–2315

147. Tareke E, Rydberg P, Karlsson P et al (2002) Analysis of acrylamide, a carcinogen formed in heated foodstuffs. J Agric Food Chem 50(17):4998–5006

148. Wolfrom ML, Kashimur N, Horton D (1974) Detection of maillard browning reaction-products as trimethylsilyl derivatives by gas-liquid-chromatography. J Agric Food Chem 22(5):791–795

149. Monti SM, Ritieni A, Graziani G et al (1999) LC/MS analysis and antioxidative efficiency of Maillard reaction products from a lactose-lysine model system. J Agric Food Chem 47 (4):1506–1513

150. Davidek T, Clety N, Devaud S et al (2003) Simultaneous quantitative analysis of Maillard reaction precursors and products by high-performance anion exchange chromatography. J Agric Food Chem 51(25):7259–7265

151. Wang J, Lu YM, Liu BZ et al (2008) Electrospray positive ionization tandem mass spectrometry of Amadori compounds. J Mass Spectrom 43(2):262–264

152. Jakas A, Katic A, Bionda N et al (2008) Glycation of a lysine-containing tetrapeptide by D-glucose and D-fructose—influence of different reaction conditions on the formation of Amadori/Heyns products. Carbohydre Resh 343(14):2475–2480

153. Frolov A, Hoffmann P, Hoffmann R (2006) Fragmentation behavior of glycated peptides derived from D-glucose, D-fructose and D-ribose in tandem mass spectrometry. J Mass Spectrom 41(11):1459–1469

154. Apweiler R, Hermjakob H, Sharon N (1999) On the frequency of protein glycosylation, as deduced from analysis of the SWISS-PROT database. Biochim Biophys Acta 1473(1):4–8

155. Yeboah FK, Alli I, Yaylayan VA (1999) Reactivities of D-glucose and D-fructose during glycation of bovine serum albumin. J Agric Food Chem 47(8):3164–3172

156. Yeboah FK, Alli I, Simpson BK et al (1999) Tryptic fragments of phaseolin from protein isolates of Phaseolus beans. Food Chem 67(2):105–112

157. Leonil J, Molle D, Fauquant J et al (1997) Characterization by ionization mass spectrometry of lactosyl beta-lactoglobulin conjugates formed during heat treatment of milk and whey and identification of one lactose-binding site. J Dairy Sci 80(10):2270–2281

158. Morgan F, Leonil J, Molle D et al (1997) Nonenzymatic lactosylation of bovine beta-lactoglobulin under mild heat treatment leads to structural heterogeneity of the glycoforms. Biochem Biophys Res Commun 236(2):413–417

159. Bouhallab S, Morgan F, Henry G et al (1999) Formation of stable covalent dimer explains the high solubility at pH 4.6 of lactose-beta-lactoglobulin conjugates heated near neutral pH. J Agric Food Chem 47(4):1489–1494

160. Morgan F, Leonil J, Molle D et al (1999) Modification of bovine beta-lactoglobulin by glycation in a powdered state or in an aqueous solution: effect on association behavior and protein conformation. J Agric Food Chem 47(1):83–91

161. Morris HR, Chalabi S, Panico M et al (2007) Glycoproteomics: past, present and future. Inter J Mass Spectrom 259(1–3):16–31

162. Smales CM, Pepper DS, James DC (2001) Evaluation of protein modification during anti-viral heat bioprocessing by electrospray ionization mass spectrometry. Rapid Commun Mass Spectrom 15(5):351–356

163. Lapolla A, Fedele D, Martano L et al (2001) Advanced glycation end products: a highly complex set of biologically relevant compounds detected by mass spectrometry. J Mass Spectrom 36(4):370–378

164. Roscic M, Versluis C, Kleinnijenhuis AJ et al (2001) The early glycation products of the Maillard reaction: mass spectrometric characterization of novel imidazolidinones derived from an opioid pentapeptide and glucose. Rapid Commun Mass Spectrom 15(12):1022–1029

165. Niwa T (2006) Mass spectrometry for the study of protein glycation in disease. Mass Spectrom Rev 25(5):713–723

166. Nakamura S, Kato A, Kobayashi K (1991) New antimicrobial characteristics of lysozyme dextran conjugate. J Agric Food Chem 39(4):647–650

167. Shu YW, Sahara S, Nakamura S et al (1996) Effects of the length of polysaccharide chains on the functional properties of the Maillard-type lysozyme-polysaccharide conjugate. J Agric Food Chem 44(9):2544–2548

168. Kato A, Nakamura S, Takasaki H et al (1996) Novel functional properties of glycosylated lysozymes constructed by chemical and genetic modifications. Macromol Interact Food Tech 650:243–256

169. Scaman C, Nakai S, Aminlari M (2006) Effect of pH, temperature and sodium bisulfite or cysteine on the level of Maillard-based conjugation of lysozyme with dextran, galactomannan and mannan. Food Chem 99(2):368–380

170. Vallejo-Cordoba B, Gonzalez-Cordova AF (2007) CE: a useful analytical tool for the characterization of Maillard reaction products in foods. Electrophoresis 28:4063–4071

171. Sanz ML, Corzo-Martinez M, Rastall RA et al (2007) Characterization and in vitro digestibility of bovine beta-lactoglobulin glycated with galactooligosaccharides. J Agric Food Chem 55(19):7916–7925

172. Kovtoun SV, English RD, Cotter RJ (2002) Mass correlated acceleration in a reflectron MALDI TOF mass spectrometer: an approach for enhanced resolution over a broad mass range. J Am Soc Mass Spectrom 13(2):135–143

173. Pettibone CJV, Schlutz FW (1916) Amino acid nitrogen in the systemic blood of children in health and disease. J Am Med Assoc 67:262–263

174. Vlassara H (2005) Advanced glycation in health and disease role of the modern environment. Maillard React 1043:452–460

175. Ritchie MA, Deery MJ, Lilley K (2002) Precursor ion scanning for structural characterization of heterogeneous glycopeptides mixtures. J Am Soc Mass Spectrom 13:1065–1077

176. Foettinger A, Leitner A, Lindner W (2006) Derivatisation of arginine residues with malondialdehyde for the analysis of peptides and protein digests by LC-ESI-MSMS. J Mass Spectrom 41(5):623–632

177. Bykova NV, Rampitsch C, Krokhin O et al (2006) Determination and characterization of site-specific N-glycosylation using MALDI-Qq-TOF tandem mass spectrometry: case study with a plant protease. Anal Chem 78(4):1093–1103

178. Cotter RJ, Griffith W, Jelinek C (2007) Tandem time-of-flight (TOF/TOF) mass spectrometry and the curved-field reflectron. J Chromatogr B 855(1):2–13

179. Liu JF, Cai Y, Wang JL et al (2007) Phosphoproteome profile of human liver Chang's cell based on 2-DE with fluorescence staining and MALDI-TOF/TOF-MS. Electrophoresis 28 (23):4348–4358

Chapter 2
Characterization of the Maillard Reaction in Amino Acid–Sugar Systems

2.1 Introduction

In order to systematically study the mechanism of the Maillard reaction, a simple platform has been built by using amino acid–sugar systems for fundamental knowledge of Amadori rearrangement products (ARPs) to predict nutritional status and assess protein quality [1–3]. In the study, MS-based techniques were employed to analyze ARPs, which provided important information about the reaction mechanisms and displayed the properties of ARPs via specific fragmentation patterns [4–8]. Results of fragmentation further provided the structural evidence, allowing identification confidently, and study of the ARPs fragmentation behavior was critical toward understanding of mechanisms for peptides and proteins in Maillard reaction.

In amino acid–sugar system, the samples were freeze-dried, purified by HPLC, and examined by quadrupole ion trap MS in the positive mode and N^{α}-acetyl-proline was selected as the internal standard for quantitative analysis. Further, the purified compounds (purity >90%) were analyzed by NMR, dissolved in dimethylsulfoxide-d_6 (DMSO-d_6, 500 µL). 1D ^{1}H NMR, ^{13}C NMR, the distortionless enhancement by polarization transfer (DEPT-135) spectrum and 2D ^{1}H–^{1}H and ^{1}H–^{13}C correlation spectroscopy (COSY) were used to assign correlations between the signals in the ^{1}H and ^{13}C NMR spectra. To confirm the connection between the amino acid and glucose, 2D nuclear Overhauser enhancement spectroscopy (NOESY) experiments were performed.

2.2 MS Analysis of Amino Acid–Sugar System

ESI-MS was used to monitor and characterize the ARPs, and samples (amino acids and sugars) were prepared at different temperatures and reaction times. The protonated amino acids of [M+H]$^+$ and [M+162+H]$^+$ (monosaccharide) were the

© The Author(s), under exclusive license to Springer Nature Switzerland AG 2018
D. Ruan et al., *The Maillard Reaction in Food Chemistry*, Chemistry of Foods
https://doi.org/10.1007/978-3-030-04777-1_2

dominant peaks. By comparing the intensities of *m/z* 189 (protonated N$^{\alpha}$-acetyl-lysine) and *m/z* 351 (protonated glycated N$^{\alpha}$-acetyl-lysine) peaks to the internal standard, the relative intensities could show the temperature effects on the formation of ARPs.

All amino acids could react with glucose noticeably by dry-heating at 50 °C, especially lysine, histidine, and arginine. In wet-heating, the reaction rate was slow. Further, ESI-MS results displayed that N-terminal acetylated lysine was only amino acids reacted with glucose and proved that the ε-NH$_2$ group was more reactive than other side-chain amino groups in Maillard reaction. Using higher reaction, temperature would shorten the reaction time of Maillard reaction. Among the three reducing sugars, glucose was the most reactive with amino compounds at relatively low temperature (50 °C). The reaction rate of sugars followed the order: glucose > lactose > maltose ≫ sucrose and raffinose. Similar results were obtained in amino acids and their N-terminal acetylated forms with other sugars (lactose, maltose, sucrose, and raffinose) (Fig. 2.1).

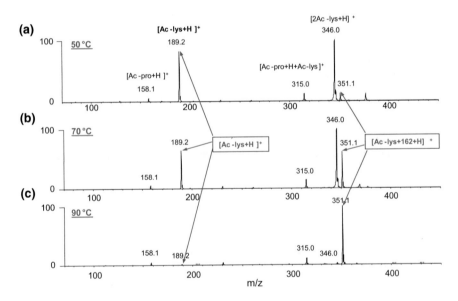

Fig. 2.1 ESI-MS spectra of glycated N$^{\alpha}$-acetyl-lysine with D-glucose dry-heating at **a** 50 °C, **b** 70 °C, and **c** 90 °C for 1 h (N$^{\alpha}$-acetyl-proline was used as internal standard)

2.3   Structure Determination of Glucosylated N$^{\alpha}$-Acetyl-Lysine by Nuclear Magnetic Resonance (NMR) Spectroscopy

A typical ARP, glucosylated N$^{\alpha}$-acetyl-lysine, was selected for further structural study. After purification by HPLC, the structure of glucosylated N$^{\alpha}$-acetyl-lysine was identified by NMR spectroscopy (Fig. 2.2); the result would facilitate interpretation of fragmentation behaviors of ARPs in subsequent MS experiments. The determination of the structure of glucosylated N$^{\alpha}$-acetyl-lysine was carried out using a series of 1D and 2D NMR spectra.

In Fig. 2.3, the ^{1}H NMR spectrum shows the two group signals of N$^{\alpha}$-acetyl-lysine moiety and typical D-glucose moiety, and the signal at 7.7 ppm suggests an amide proton (3-NH) of lysine moiety, and this proton is strongly correlated with the proton at 4.0 ppm according to the 2D ^{1}H–^{1}H COSY spectrum (Fig. 2.4). Meanwhile, the correlated carbon of the proton at 4.0 ppm is at 51 ppm

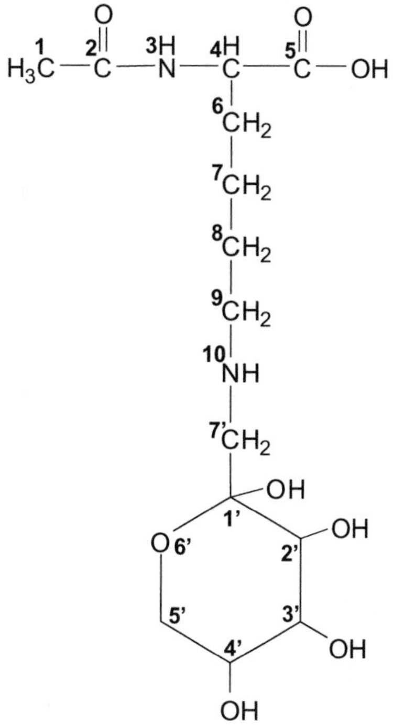

Fig. 2.2   Structure of glucosylated N$^{\alpha}$-acetyl-lysine

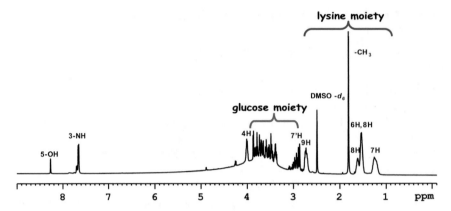

Fig. 2.3 ^{1}H NMR spectrum of glucosylated N^α-acetyl-lysine (the assignments of the protons are according to the numerals in Fig. 2.5)

Fig. 2.4 ^{1}H–^{1}H COSY spectrum of glucosylated N^α-acetyl-lysine (the assignments of the cross-peaks are according to the numerals in Fig. 2.2)

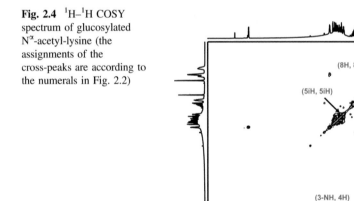

(Fig. 2.5), which is assigned to a –CH group by DEPT-135 spectrum. Based on this information, the proton at 4.0 ppm could be verified as α-H (4-H) of lysine moiety. The ^{1}H–^{1}H COSY spectrum also shows the other correlations between the protons at 4.0 and 1.5 ppm, 1.5 and 1.3 ppm, 1.5 and 2.7 ppm, indicating the connection of lysine alphabetic chain from α-H (Fig. 2.4). According to the integral of the signals, the overlapping of 6-H and 8-H at 1.5 ppm could be observed. The correlations between the protons at 1.6 and 1.5 ppm, 1.6 and 2.7 ppm suggest that the signals could be assigned to the two protons at position 8. The distortionless enhancement by polarization transfer (DEPT-135) spectrum is a spectral editing sequence, which could discriminate –CH, –CH$_2$ and –CH$_3$ carbons. In the –CH$_2$ profile, signals at

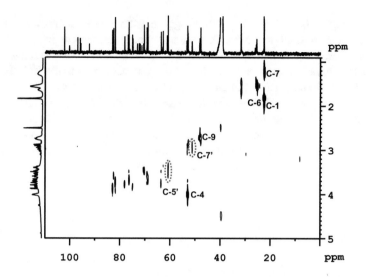

Fig. 2.5 ^{13}C–^1H COSY spectrum of glucosylated N^α-acetyl-lysine (the assignments of the cross-peaks are according to the numerals in Fig. 2.2)

Fig. 2.6 2D NOESY spectrum of glucosylated N^α-acetyl-lysine (the assignments of the cross-peaks are according to the numerals in Fig. 2.2)

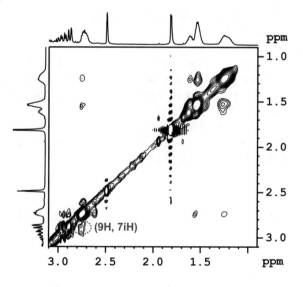

26, 22, 32, 48 ppm, which are correlated to these protons in ^{13}C–^1H COSY spectrum (Fig. 2.5), also indicate that they are β-C to δ carbon of lysine moiety. The ^{13}C–^1H COSY spectrum suggests that the protons at position 8 split at 1.6 and 1.5 ppm, which is also supported by the strong correlation with ^1H–^1H COSY spectrum (Fig. 2.6).

The complexity of the NMR spectra suggests that the glucose moiety might exist in different conformations. Nevertheless, the strong correlations in ^1H-^1H COSY spectrum between two doublets at 3.4 and 3.5 ppm could be observed, and the two signals are connected to the same carbon at 61 ppm according to the ^{13}C-^1H COSY spectrum. The two signals at 3.4 and 3.5 ppm were then assigned to 5'-H of glucose moiety. Moreover, the 7'-H also could be assigned based on the NMR data, which is important for the crosslink between the two moieties. A doublet peak at 2.9 ppm was observed, and their correlated carbon at 51 ppm was $-CH_2$ profile in the DEPT-135 spectrum. In the NOESY spectrum (Fig. 2.6), the proton had a NOE effect with proton at position 9, which verified the connection between lysine and glucose moieties.

2.4   ARPs in Amino Acid–Sugar Systems

2.4.1   MS/MS of ARPs with D-Glucose

In the MS/MS spectra, dominant fragmentation ions come from the neutral loss of molecule(s) of water by 18 Da ($-H_2O$), 36 Da ($-2H_2O$), 54 Da ($-3H_2O$), 84 Da ($-3H_2O-HCOH$) and/or whole glucose moiety by loss of 162 Da. All these losses are cleaved from the sugar moiety in MS/MS experiments as previously reported [6, 9, 10].

In Fig. 2.7, the dominant fragment ion is $[M-18+H]^+$ by loss of one water molecule ($-H_2O$) from the protonated ARPs, and the second strongest peak is $[M-36+H]^+$ by loss of two H_2O molecules. The characteristic fragmentation patterns of ARPs in the amino acid–glucose systems suggest that sugar moieties were preferentially cleaved in the low collision energy. For the MS/MS spectra of glucosylated (N^α-acetyl-) lysine and glutamine, the typical ions of $[M-84+H]^+$ ($-3H_2O-HCOH$) are shown in Fig. 2.7c–e.

2.4.2   MS/MS of ARPs with Disaccharides

MS/MS spectra of ARPs in amino acid–disaccharide systems show the main fragmentation ions of ARPs by the loss of one and two molecules of water from the sugar moieties. It is similar to the ARPs of amino acid–glucose. When lysine reacts with lactose and maltose, the cleavage of whole sugar moiety occurred with a mass loss of m/z 162 and m/z 324. The fragmentation ions of maltose and lactose moieties are too similar to be differentiated from the MS/MS data [11, 12]. Among the fragmentation ions, the m/z 225 was observed in MS/MS, suggesting that the basic skeleton of the disaccharides is similar to glucose. Theoretically, the cleavage of

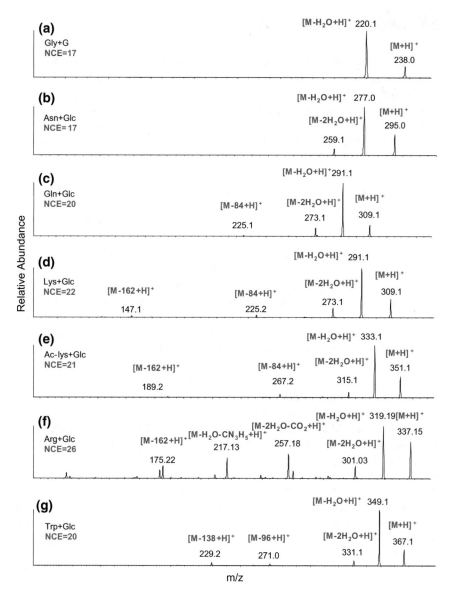

Fig. 2.7 MS/MS spectra of glycated **a** glycine, **b** asparagine, **c** glutamine, **d** lysine, **e** N^α-acetyl-lysine, **f** arginine, and **g** tryptophan

one monosaccharide moiety (−162 Da) is followed by loss of 84 Da ($3H_2O+HCHO$), verified in further MS^3 experiments. The peaks of m/z 225 from different ARPs of lysine reacting with sugars show similar fragmentation ions in MS^3 experiments.

2.5 MS/MS and MS^3 of ARPs

The dominant ions of APRs in MS/MS are fragments of losing 1–3 H_2O molecules, a HCHO molecule and glucose/disaccharide moiety, as $[M-18+H]^+$, $[M-36+H]^+$, $[M-54+H]^+$, $[M-84+H]^+$, and/or $[M-n \times 162+H]^+$, and these fragment ions were formed by neutral losses of formaldehyde or water [10–14]. Since $[M-84+H]^+$ is one of the important ions, forming a 5-membered ring ion from sugar moiety, it has been well characterized in the studies, which having different forms of $[M-84+H]^+$ [11–17]. In our study, the ε-NH_2 group of lysine participates in the Maillard reaction and has been shown to be the target group in peptides and proteins that react with reducing sugars. Hence, the study of lysine and N$^\alpha$-acetyl-lysine fragmentation pattern should help to understand the cleavage behavior of glycated peptides and proteins. In the MS/MS spectrum of N$^\alpha$-acetyl-lysine, m/z 171 is by loss of H_2O and m/z 129 is immonium ion of lysine. It implied that the α-NH_2 group is more fragile than the ε-NH_2 group under the same collision energy and m/z 129 is more specific without free N-terminal amino group. This property might provide additional information for identifying lysine in peptide and protein mapping.

In MS^3 spectrum of m/z 130, m/z 84 is the only fragment, indicating that m/z 101 could not be produced from m/z 130, probably due to fragmentation mechanism forming different structures. Based on the MS/MS spectra, the fragmentation pathway of lysine and N$^\alpha$-acetyl-lysine was postulated. In order to obtain more information on the $[M-84+H]^+$ peaks from ARPs, the specific fragment was isolated and further fragmented in MS^3. The peaks of m/z 225 were ions of $[M-84+H]^+$ from ARPs of lysine reacting with glucose, maltose, and lactose. Although they origi- nated from different ARPs of lysine reacting with different sugars in the MS/MS experiments, their MS^3 spectra show no discernible difference. This suggests that the three sugars have similar skeleton of oxonium ions. The peaks of m/z 267 from N$^\alpha$-acetyl-lysine reacting with different sugars show the same results. MS^3 spectra of m/z 225 and m/z 267 show the cleavage of amino acid moiety by loss of 44 Da ($-CO_2$), 46 Da ($-HCO_2H$), and some specific ions of lysine, such as ions of m/z 130 and m/z 84. MS^3 of these typical ions provide the structural information and reaction sites of lysine [12, 14]. All fragmentation ions in MS^3 spectra were assigned (Fig. 2.8). The data provide helpful information for subsequent studies of glycation sites of peptides and proteins.

2.6 Summary

The amino acid–sugar systems used in the present investigation demonstrated the effects of different variables such as the type of amino acids and sugars, tempera- ture, and time on the rate of the Maillard reaction. Dry-heating was preferred over wet-heating and increased temperature correlated with higher reaction rate. Positively charged amino acids preferentially reacted with reducing sugars. Besides,

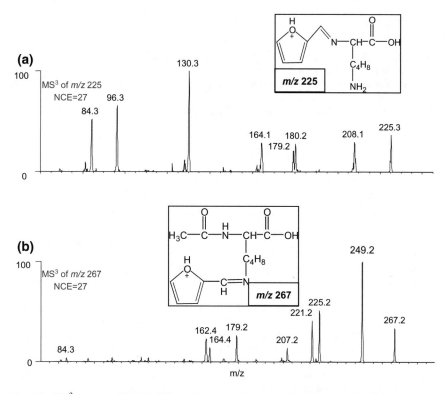

Fig. 2.8 MS^3 spectra of [M-84+H]$^+$: **a** glucosylated lysine and **b** glucosylated N$^\alpha$-acetyl-lysine

the primary amino group of the amino acids, the ε-amino group of lysine also showed high reactivity.

The fragmentation behaviors of the ARPs showed that the sugar moiety tended to be fragmented preferentially by neutral loss of small molecules to form oxonium ions. The fragmentation behaviors of lysine and N$^\alpha$-acetyl-lysine were studied, and the cyclical mechanisms were proposed based on MS/MS and MS^3 experiments. One specific ARPs of N$^\alpha$-acetyl-lysine with glucose was isolated by HPLC, and its structure was characterized by NMR spectroscopy. In MS^3 spectra, the peaks of [M-84+H]$^+$ were further fragmented and the data show that the skeletons of different sugars (oxonium ions) appear to be the same.

Computational results were consistent with experimental observations, supporting the proposed ARPs fragmentation mechanism. The fragmentation study in gas phase could serve as the basis for further studies on complex systems, such as the use of lysine-containing peptides and the identification of protein glycation sites, to be presented in following chapters.

References

1. Machiels D, Istasse L (2002) Maillard reaction: importance and applications in food chemistry. Ann Med Veter 146(6):347–352
2. Friedman M (1996) Food browning and its prevention: an overview. J Agric Food Chem 44 (3):631–653
3. Finot PA (2005) Historical perspective of the Maillard reaction in food science. Maillard React 1043:1–8
4. Yalcin T, Csizmadia IG, Peterson MR et al (1996) The structure and fragmentation of B-n (n >= 3) ions in peptide spectra. J Am Soc Mass Spectrom 7(3):233–242
5. Yalcin T, Khouw C, Csizmadia IG et al (1995) Why are B ions stable species in peptide spectra? J Am Soc Mass Spectrom 6(12):1165–1174
6. Mennella C, Visciano M, Napolitano A et al (2006) Glycation of lysine-containing dipeptides. J Pep Sci 12(4):291–296
7. Borrelli RC, Visconti A, Mennella C et al (2002) Chemical characterization and antioxidant properties of coffee melanoidins. J Agric Food Chem 50(22):6527–6533
8. Cotham WE, Hinton DJS, Metz TO et al (2003) Mass spectrometric analysis of glucose-modified ribonuclease. Biochem Soc Trans 31:1426–1427
9. Wrodnigg TM, Eder B (2001) The Amadori and Heyns rearrangements: landmarks in the history of carbohydrate chemistry or unrecognized synthetic opportunities? Glyoscience 215:115–152
10. Wang J, Lu YM, Liu BZ et al (2008) Electrospray positive ionization tandem mass spectrometry of Amadori compounds. J Mass Spectrom 43(2):262–264
11. Li C, Wang H, Zhang YY et al (2014) Characteristics of early maillard reaction products by electrospray ionization mass spectrometry. Asian J Chem 26(21):7452–7456
12. Zhang YY, Ruan ED, Wang H et al (2014) A fundamental study of Amadori rearrangement products in reducing sugar-amino acid model system by electrospray ionization mass spectrometry and computation. Asian J Chem 26(10):2914–2944
13. Horvat S, Jakas A (2004) Peptide and amino acid glycation: new insights into the Maillard reaction. J Pep Sci 10(3):119–137
14. Ruan ED, Wang H, Ruan YY et al (2013) Study of fragmentation behavior of Amadori rearrangement products in lysine-containing peptide model by tandem mass spectrometry Eur. J Mass Spectrom 19(4):295–303
15. Taylor VF, March RE, Longerich HP et al (2005) A mass spectrometric study of glucose, sucrose, and fructose using an inductively coupled plasma and electrospray ionization. Inter J Mass Spectrom 243(1):71–84
16. Jakas A, Katic A, Bionda N et al (2008) Glycation of a lysine-containing tetrapeptide by D-glucose and D-fructose—influence of different reaction conditions on the formation of Amadori/Heyns products. Carbohyd Res 343(14):2475–2480
17. Jeric I, Versluis C, Horvat S et al (2002) Tracing glycoprotein structures: electron ionization tandem mass spectrometric analysis of sugar-peptide adducts. J Mass Spectrom 37(8):803–811

Chapter 3
Characterization of Glycated Lysine in Peptide–Sugar System

3.1 Introduction

The model of lysine–reducing sugars had been studied by Maillard [1] and many other researchers [2–4]. However, free amino acids are present only in some vegetables and fruits and most of the ARPs are produced on the side chain of lysine residues present in peptides and proteins both in foods and in vivo [5]. In the last twenty years, glycation had been one of the most widely spread non-enzymatic side-chain-specific posttranslational modifications (PTMs) by the Maillard reaction and it produced many derivatives from peptides and proteins with free amino functional groups, such as ε-amino group of lysine residue [6–8]. MS and MS-related techniques had become more and more useful and powerful for analyzing the structure and characterizing posttranslational modifications of glycated peptides and proteins. Many researches had been quantified ARPs in model protein systems and in foods. For example, studies conducted on N-terminal Amadori modified model peptides using ESI-MS revealed characteristic fragmentation patterns [9, 10]. Iberg and Fluckiger [11] observed that in vivo glycosylation of human serum albumin mainly occurred on lysine residues with the near-neighbor effects of another close amino group. Venkatraman and Chan [12] also studied the near-neighbor effects on the reactivity of amino group on the side chains. Further, techniques reported for ARPs were selected ions monitoring [13], multiple reaction monitoring [14], and neutral loss scan [15, 16]. Studies showed that the consecutive losses of one or three water molecules with an additional molecule of formaldehyde (−84 Da) could yield oxonium, pyrylium, and furylium ions [9, 17, 18]. The results emphasized the structural determination of Schiff base and monitoring of ARPs formation to show the important role of the unique ε-amino group of lysine in modification of foodstuff. Thus, the derivatives like ARPs of lysine-containing peptides and proteins with reducing sugars in the Maillard reaction were widely used to evaluate the extent of the early stage of the Maillard reaction in food [10, 19, 20].

© The Author(s), under exclusive license to Springer Nature Switzerland AG 2018
D. Ruan et al., *The Maillard Reaction in Food Chemistry*, Chemistry of Foods
https://doi.org/10.1007/978-3-030-04777-1_3

In this investigation, the fragmentation behavior and the structures of specific ARPs of peptide–sugar systems were studied, and MS techniques were employed to monitor the glycation of lysine-containing peptides and to evaluate the effects of reactants, reaction conditions, and position of lysine on the Maillard reaction. Further, to study the effect of the position of lysine on glycation, lysine-containing peptides were designed and synthesized by altering the lysine position in a series of peptides. Fragmentation behaviors of ARPs of lysine-containing peptides were also characterized, and the corresponding glycation sites were identified by MS/MS experiments. In the peptide–sugar model, peptide–sugar linkages were chosen to figure out the differences in the fragmentation pattern and to identify characteristic fragments suitable for the further study in complex systems.

3.2 Solid-Phase Peptide Synthesis (SPPS)

3.2.1 Synthesis Procedure

Loading: Fmoc-protected amino acid was dissolved in distilled dichloromethane (DCM) and dimethylformamide (DMF). 1,3-Diisopropylcarbodiimide (DIC) was added to the Fmoc-protected amino acid mixture on ice for 30 min to generate a reactive intermediate. Wang resin was swelled in DMF for 30 min, filtered, washed with DMF and DCM, and then dried under nitrogen gas in fritted reaction flask. The reactive intermediate was completely dissolved in DMF and transferred to the pre-washed Wang resin. N,N-Dimethyl 4-aminopyridine (DMAP) was added to the reaction mixture. The mixture was agitated for 2 h with a flow of nitrogen at room temperature. The resin was filtered, washed with DMF (3 times), DCM (3 times), and dried under nitrogen gas. The optimized Wang resin loading ranged from 0.8 to 1.2 mmol/g for peptide synthesis.

DeFmoc: Freshly prepared 20% piperidine in DMF (v/v) was added to the Wang resin with Fmoc-protected amino acid. After the piperidine was attached to the resin at room temperature, the mixture was swirled for 20 min by nitrogen bubbling. The prepared resin was filtered, washed with DMF and DCM, and then dried under nitrogen gas.

Coupling: Fmoc-amino acid dissolved in DMF was added to the DeFmoc Wang resin bonding amino acid with 1 M 1-hydroxybenzotriazole hydrate (HOBT) in DMF and DIC in DMF. The reaction mixture was swirled under nitrogen gas for 1.5 h at room temperature by nitrogen bubbling. The resin was filtered, washed with DMF and DCM, and then dried under nitrogen gas. The desired amino acid sequence could be achieved by repeating the **DeFmoc** and **Coupling** procedures.

3.2.2   Cleavage of the Peptide

DeFmoc: Wang resin with Fmoc-protected peptide was prepared as described above.

   Cleavage: The Wang resin with peptide attached was swirled by nitrogen bubbling in the presence of a cleavage solution for 2 h at room temperature. Cleavage solution was prepared with trifluoroacetic acid (TFA), DCM, tris-isopropylsilane (TIS) and deionized water (in ratio v/v/v/v, TFA:DCM:TIS:H_2O = 70:20:5:5). The resin was filtered and washed with TFA. The filtrate solution was combined and reduced to about 5 mL under vacuum bump and was transferred to tert-butyl methyl ether, kept at −20 °C overnight. The precipitate was collected by centrifugation and washed with tert-butyl methyl ether and completely dried by vacuum overnight.

3.3   ESI-MS Analysis

The concentration of amino acids was 0.1 M, and the concentration of sugars was 1.0 M. Each amino acid was mixed with each sugar at 1:1 (v/v). All samples were freeze-dried and dry-heated at 50, 70, and 90 °C, respectively. Then samples were kept at −20 °C immediately after reaction before analysis.

   All mass spectrometric experiments were conducted in the positive mode with a quadrupole ion trap mass spectrometer, LCQ Deca XP Plus (Finnigan LCQ, Thermo Finnigan, San Jose, CA, USA) equipped with a nano-spray ion source. Ion spray tips were pulled by using 150 μm (OD) × 50 μm (ID) capillary on fire. Electrospray voltage was typically kept between 2.8 and 3.0 kV, and the inlet capillary was maintained at 180 °C. To obtain the spectra of MS/MS or MS^n, the normalized collision energy was varied with all the other ion tuning conditions fixed. Samples typically comprised of 1.0 μM peptides in a water/methanol (50:50) solution with 1% formic acid, and the physical parameters of the interface were optimized at the flow rate of 25 μL/h, i.e., the distance of the needle from the hole in the spray shield (1.2–1.5 cm), the high voltage (2.8–3 kV) added on the stainless steel unit and the temperature of the heated capillary (180 °C). The samples of peptides and glycated peptides under CID conditions were directly infused at a nano-rate of 25 μL/h for analysis in positive ionization mode continuously and were acquired using helium as the collision gas. The injection and activation times for CID in the ion trap were 200 and 30 ms, respectively.

3.4   Effects of the Maillard Reaction

The effects of sugars, reaction temperature, and time on the rate of the Maillard reaction in peptide–sugar systems were similar to those observed in the amino acid–sugar systems. Glucose could react with all lysine-containing peptides readily at

50 °C, while the reactivity of lactose and maltose with peptides was poor or even undetectable by MS at 50 °C for 1 h. When the reaction temperature was increased to 90 °C, all the three reducing sugars could react with peptides, showing that the sugar effect on the rate of ARP production decreased as the temperature increased. Figure 3.1 shows the temperature effect on the rate of Maillard reaction between D-glucose and (N^{α}-acetyl-) lysine and di-peptides.

Based on the MS data, the Maillard reaction rate in peptide–sugar systems decreased as the length of peptides increased, and the reaction rates of di-peptides were higher than that of tri-peptides at 50 °C. This may be due to steric hindrance affecting the rate of Maillard reaction at relatively low temperature. However, the effect of peptide length was gradually reduced with increasing reaction temperature. Figure 3.2 shows that the peptide length and lysine position have no noticeable effects on the rate of Maillard reaction at 70 °C. Di-peptides of KM and MK were more reactive than lysine and acetyl-lysine, and the tetra-peptide (TYSK) had reaction rate close to those of three tri-peptides (KTY, KYS, and YSK).

3.5 Fragmentation Sugar Moieties

Fragmentation behavior of the ARPs derived from lysine-containing peptides reacting with reducing sugars was characterized by MS/MS. The most abundant fragments obtained from ARPs of glucosylated peptides were constituted by the ions of $[M-18+H]^{+}$, $[M-36+H]^{+}$, $[M-54+H]^{+}$, $[M-72+H]^{+}$, $[M-84+H]^{+}$, and

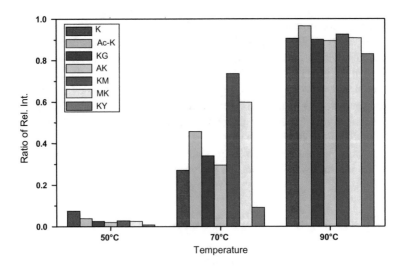

Fig. 3.1 Ratio of relative intensity of ARPs of (N^{α}-acetyl-) lysine and di-peptides reacting with D-glucose at three different temperatures under dry-heating for 1 h

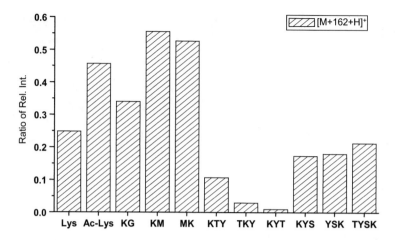

Fig. 3.2 Ratio of relative intensity of glucosylated lysine derivatives and lysine-containing peptides under dried heated at 70 °C for 1 h

$[M-162+H]^+$ (Fig. 3.3), due to the loss of one to four molecules of waters, HCHO and glucose from the glucose moiety, respectively [21]. In Tables 3.1 and 3.2, the most abundant fragments obtained from ARPs of di-peptides and KxK series tri-peptides reacting with glucose are listed. Data show that the ARPs formed from different peptides exhibited a common and characteristic fragmentation pattern constituted by typical ions from the sugar moiety. The fragmentation pattern is also in good agreement with previous reports [8, 22, 23].

For ARPs of peptides reacting with disaccharides (Fig. 3.4), the fragments in MS/MS spectra showed a similar fragmentation pattern constituted mainly by the ions of $[M-18+H]^+$, $[M-36+H]^+$, $[M-246+H]^+$, and $[M-324+H]^+$, due to the loss of one and two molecules of waters, partial sugar moiety, and whole sugar moiety, respectively (Fig. 3.4).

In further MS/MS study of ARPs of glucosylated lysine-containing peptides, two peak clusters were observed clearly in Fig. 3.5b–d. Peak clusters at high mass range ($m/z > 350$) were typical fragmentation ions from the sugar moiety by loss of one to three water molecules and by loss of 84 Da ($3H_2O+HCHO$). Other peak cluster at low mass range (m/z range at 200–300) was constituted of specific fragment ions of m/z 291, m/z 273, m/z 255, m/z 225, and/or m/z 207, which were also observed from the MS/MS spectrum of ARPs of lysine reacting with glucose (Fig. 3.5a). The data suggest that glucose is probably covalently bonded with lysine in these peptides through the Maillard reaction. Based on the fragmentation mechanism of glucose moiety mentioned previously [24, 25], the specific product ions of glycated peptides related were assigned to the fragments derived from lysine residues: m/z 291, $[K^*-H_2O+H]^+$; m/z 273, $[K^*-2H_2O+H]^+$, m/z 255, $[K^*-3H_2O+H]^+$, m/z 225, $[K^*-3H_2O-HCHO+H]^+$, and m/z 207, $[K^*-4H_2O-HCHO+H]^+$, respectively, (among $K^*=K+Glc$) (Fig. 3.5).

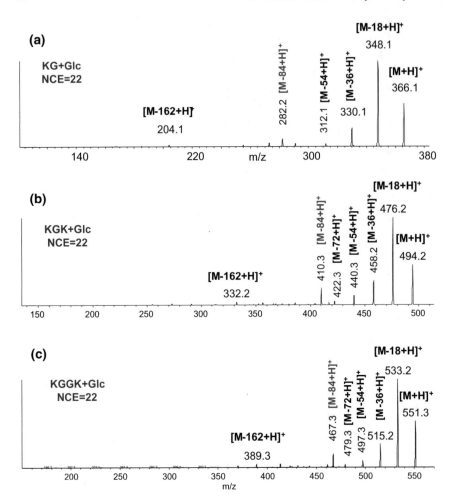

Fig. 3.3 MS/MS spectra of ARPs of: **a** KG, **b** KGK, and **c** KGGK reacting with glucose

Generally, the fragmentation of ARPs produced an oxonium ion of sugar formed by the loss of water and the cleavage of anomeric hydroxyl group [26]. The fragmentation patterns controlled by the sugar and peptide moiety were a reflection of the structural characteristics of the ARPs. Among the fragments, we found the specific ions by loss of 84 Da ($-3H_2O$–HCHO) from glucose moiety or 246 Da ($-C_6H_{12}O_5$–$3H_2O$–HCHO) from disaccharide moiety. For example, the peaks of m/z 282, corresponding to specific mass loss, were observed in MS/MS spectra of ARPs from di-peptide (KG) reacting with different sugars (Figs. 3.3a and 3.4a, b), and the peaks of m/z 410 were also observed in MS/MS spectra of ARPs from KGK reacting with glucose (Fig. 3.3b), lactose (Fig. 3.4c), and maltose (Fig. 3.4d). The ions of [M-84+H]$^+$/[M-246+H]$^+$ were produced by forming the same skeleton of

Table 3.1 Main fragmentation ions in MS/MS spectra of ARPs of di-peptides reacting with glucose (Glc represents glucose)

Peptide	ARPs	$[M+H]^+$ m/z	MS/MS ions m/z	Loss	Fragment lost
KG	KG-Glc	366	348	−18	–H_2O
			330	−36	–$2H_2O$
			312	−54	–$3H_2O$
			282	−84	–$3H_2O$–HCHO
			204	−162	$C_6H_{10}O_5$
GK	GK-Glc	366	348	−18	–H_2O
			330	−36	–$2H_2O$
			312	−54	–$3H_2O$
			282	−84	–$3H_2O$–HCHO
			204	−162	$C_6H_{10}O_5$
KM	KM-Glc	440	422	−18	–H_2O
			404	−36	–$2H_2O$
			386	−54	–$3H_2O$
			356	−84	–$3H_2O$–HCHO
			278	−162	$C_6H_{10}O_5$
MK	MK-Glc	440	422	−18	–H_2O
			404	−36	–$2H_2O$
			386	−54	–$3H_2O$
			356	−84	–$3H_2O$–HCHO
			278	−162	$C_6H_{10}O_5$
AK	AK-Glc	380	362	−18	–H_2O
			344	−36	–$2H_2O$
			326	−54	–$3H_2O$
			296	−84	–$3H_2O$–HCHO
			278	−162	$C_6H_{10}O_5$
KY	KY-Glc	472	454	−18	–H_2O
			436	−36	–$2H_2O$
			418	−54	–$3H_2O$
			388	−84	–$3H_2O$–HCHO
			310	−162	$C_6H_{10}O_5$

oxonium ions derived from monosaccharides and disaccharides, respectively. These ions were also detected in the majority of the ARPs studied and were investigated using high-energy collision-induced dissociation (Roscic et al. 2001; Jeric et al. 2002).

Furthermore, the fragmentation patterns of the ARPs indicate that the ε-amino group of lysine related preferentially with sugar over the primary amino group at N-terminal (i.e., YSK) in the Maillard reaction. Theoretically, α-NH_2 of tyrosine and ε-NH_2 of lysine in YSK could react with glucose. However, the MS/MS spectrum of glucosylated YSK shows the typical series of ions at m/z 291, m/z 273,

Table 3.2 Main fragmentation ions in MS/MS spectra of ARPs of tri-peptides (K × K series) reacting with glucose (Glc represents glucose)

Peptide	ARPs	[M+H]$^+$ m/z	MS/MS ions m/z	Loss	Fragment lost
KGK	KGK-Glc	494	476	−18	–H_2O
			458	−36	–2H_2O
			440	−54	–3H_2O
			410	−84	–3H_2O–HCHO
			332	−162	$C_6H_{10}O_5$
KFK	KFK-Glc	584	566	−18	–H_2O
			548	−36	–2H_2O
			530	−54	–3H_2O
			500	−84	–3H_2O–HCHO
			422	−162	$C_6H_{10}O_5$
KLK	KLK-Glc	550	532	−18	–H_2O
			514	−36	–2H_2O
			496	−54	–3H_2O
			466	−84	–3H_2O–HCHO
			388	−162	$C_6H_{10}O_5$
KQK	KQK-Glc	565	547	−18	–H_2O
			529	−36	–2H_2O
			511	−54	–3H_2O
			481	−84	–3H_2O–HCHO
			403	−162	$C_6H_{10}O_5$
KYK	KYK-Glc	600	582	−18	–H_2O
			564	−36	–2H_2O
			546	−54	–3H_2O
			516	−84	–3H_2O–HCHO
			438	−162	$C_6H_{10}O_5$
KKK	KKK-Glc	565	547	−18	–H_2O
			529	−36	–2H_2O
			511	−54	–3H_2O
			481	−84	–3H_2O–HCHO
			403	−162	$C_6H_{10}O_5$

m/z 255, and m/z 225 (Fig. 3.5d), suggesting that glucose reacted with ε-NH_2 of lysine. The glycation sites could be further identified through MS^3 experiments of [M-84+H]$^+$.

Fragmentation behavior of ARPs of lysine-containing peptides reacting with lactose and maltose was also studied by MS/MS experiments. Compared to glucose, disaccharides have higher molecular weight and can make the modification easier to be detected by molecular weight-related techniques [27]. Comparatively, lactose was one of the common disaccharides used to study in the Maillard reaction due to its relatively high reactivity upon the reaction [28–31]. Similarly, the major

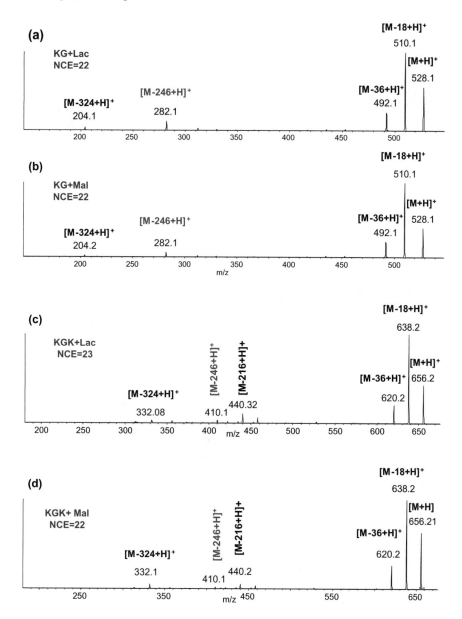

Fig. 3.4 MS/MS spectra of ARPs of: **a** KG and **c** KGK reacting with lactose (Lac), and **b** KG and **d** KGK reacting with maltose (Mal)

fragments were cleaved from the sugar moiety by the loss of one and two molecules of water and even the whole sugar moiety. The characteristic losses of 216, 246, and 324 Da were observed in the ARPs of peptides reacting with disaccharides. These ions were attributed to the loss of the galactose moiety plus three water

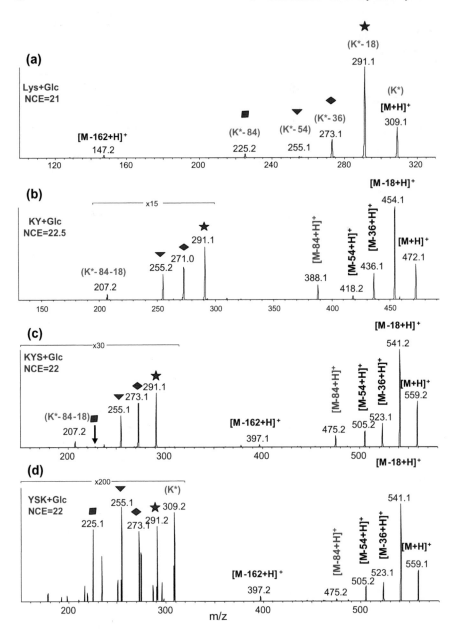

Fig. 3.5 MS/MS spectra of the glucosylated **a** K*, **b** K*G, **c** K*YS, and **d** YSK*, in which * represents glucose adducted to lysine. (A full star (filled star: –H₂O), full diamond (filled diamond: –2H₂O), full triangle (filled inverted triangle: –3H₂O), and full square (filled square: –3H₂O–HCHO) of glucosylated lysine)

molecules (216 Da), the galactose moiety plus three water molecules and one formaldehyde molecule (246 Da), and the whole sugar moiety (324 Da), respectively. In Table 3.3, the most abundant fragments obtained from ARPs formed by peptides reacting with disaccharides (maltose and lactose) are displayed. Data showed that the ARPs formed from different peptides possessed a common characteristic fragmentation pattern containing typical ions from the sugar moiety. Since the MS/MS spectra of ARPs of peptides reacting with disaccharides were almost identical to those in the peptide–sugar systems, it was hard to differentiate disaccharides using only MS/MS experiments (Fig. 3.4; Table 3.3).

3.6 Fragmentation of Peptide Moieties

MS/MS is a powerful method to identify the sequence of peptides, and the product ions of peptides are named as series of a-, b-, and y-ions. Figure 3.6 shows the normal product ions in MS/MS experiment of ion trap ESI-MS, in which N-terminal ions are kept in a- and b-ion, and C-terminal ions are kept in y-ion [32].

MS/MS experiments have the capacity to obtain enough sequence information for tri- or longer peptides. MS/MS spectra provide the entire y-series ions and partial a-/b-series ions of YSK and KYS (Fig. 3.7). It was enough to differentiate the two tri-peptides (YSK and KYS) according to the different y-/b-series ions [21]. For example, y_2-ion of peptide of YSK was m/z 234.2 and b_2-ion was m/z 251.0 (Fig. 3.7a); but y_2-ion of peptide of KYS was m/z 269.1 and b_2-ion was m/z 292.1 (Fig. 3.7b).

3.6.1 Fragmentation of $[M-84+H]^+$

One of the properties of MS/MS experiments of ion trap ESI-MS is that it could break the weak bond without specific function. Therefore, MS/MS spectra of glycated peptides displayed the product ions from the sugar moiety due to the loss of water from hydroxyl group in sugars. To obtain the peptide sequence information and to identify the glycation sites, further MS/MS experiments of specific product ions were required. Selected ions should mainly arise from the cleavage at the peptide bonds and should be relatively stable for isolation by MS instrument. Among the product ions of ARPs in peptide–sugar systems, the peaks of $[M-84+H]^+$ of glucose and $[M-246+H]^+$ of disaccharides were selected to obtain more sequence information and to identify the glycation sites. These ions had relatively stable sugar skeleton of oxonium ions, and the corresponding sugar moiety was not readily fragmented further. Moreover, the product ions could be produced with relatively high intensity in the MS/MS experiments and could be isolated for further MS/MS experiments [21]. To employ collision energy on these ions, the typical fragmentation ions of peptide-bond broken were observed, such as the typical

Table 3.3 Main fragmentation ions in MS/MS spectra of ARPs from peptides reacting with lactose and maltose (Lac: lactose; Mal: maltose)

Peptide	ARPs	$[M +H]^+$ m/z	MS/MS ions m/z	Loss	Fragment lost
KG	KG-Lac	528	510	−18	–H_2O
			492	−36	–$2H_2O$
			312	−216	–$C_6H_{10}O_5$–$3H_2O$
			282	−246	–$C_6H_{10}O_5$–$3H_2O$–HCHO
			204	−324	–$2C_6H_{10}O_5$
KGK	KGK-Lac	656	638	−18	–H_2O
			620	−36	–$2H_2O$
			440	−216	–$C_6H_{10}O_5$–$3H_2O$
			410	−246	–$C_6H_{10}O_5$–$3H_2O$–HCHO
			332	−324	–$2C_6H_{10}O_5$
KGGK	KGGK-Lac	713	695	−18	–H_2O
			677	−36	–$2H_2O$
			515	−216	–$C_6H_{10}O_5$–$3H_2O$
			497	−246	–$C_6H_{10}O_5$–$3H_2O$–HCHO
			389	−324	–$2C_6H_{10}O_5$
KG	KG-Mal	528	510	−18	–H_2O
			492	−36	–$2H_2O$
			312	−216	–$C_6H_{10}O_5$–$3H_2O$
			282	−246	–$C_6H_{10}O_5$–$3H_2O$–HCHO
			204	−324	–$2C_6H_{10}O_5$
KGK	KGK-Mal	656	638	−18	–H_2O
			620	−36	–$2H_2O$
			440	−216	–$C_6H_{10}O_5$–$3H_2O$
			410	−246	–$C_6H_{10}O_5$–$3H_2O$–HCHO
			332	−324	–$2C_6H_{10}O_5$
KGGK	KGGK-Mal	713	695	−18	–H_2O
			677	−36	–$2H_2O$
			515	−216	–$C_6H_{10}O_5$–$3H_2O$
			497	−246	–$C_6H_{10}O_5$–$3H_2O$–HCHO
			389	−324	–$2C_6H_{10}O_5$

peptide fragments of y-, b-, and a-series ions, the loss of 17 Da (–NH_3), 28 Da (–CO), 35 Da (–H_2O–NH_3), 44 Da (–CO_2) and even related immonium ions of amino acids binding with sugar skeletons.

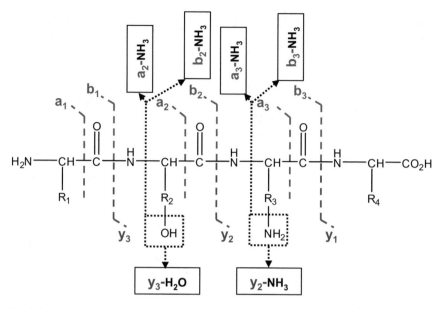

Fig. 3.6 Typical fragmentation patterns in the tandem mass spectrometry

3.6.2 *MS³ of Glucosylated Di-peptides*

MS^3 spectra of $[M-84+H]^+$ of glycated di-peptides could provide the information to identify the glycation sites. Figure 3.8 shows the typical product ions of MS^3 of $[M-84+H]^+$ of ARPs derived from di-peptides reacting with glucose. In these spectra, several typical product ions of glycated peptides related to lysine residues were observed and clearly showed the glycation site of lysine in the peptides, such as *m/z* 225 ($[K^*-84+H]^+$), *m/z* 207 ($[K^*-84-H_2O+H]^+$), *m/z* 179 ($[K^*-84-HCO_2H+H]^+$), and *m/z* 162 ($[K^*-84-HCO_2H-NH_3+H]^+$), respectively, (K^*=K+Glc). Based on the fragmentation data of MS^3 experiments of $[M-84+H]^+$, it was confirmed that lysine preferentially reacted with sugars in the Maillard reaction, whether it was at N-terminal (Fig. 3.8a, c) or C-terminal (3.8B). On the other hand, for di-peptides reacting with disaccharides, MS^3 spectra of $[M-246+H]^+$ show very similar fragmentation behavior of the specific ion of $[M-84+H]^+$.

Furthermore, from the MS data of the peptide moiety, the peaks of *m/z* 207 ($[K^*-84-H_2O+H]^+$), *m/z* 179 ($[K^*-84-HCO_2H+H]^+$) and *m/z* 162 ($[K^*-84-HCO_2H-NH_3+H]^+$) could be assigned to the immonium ions of lysine *m/z* 129, *m/z* 101, and *m/z* 84, respectively. Based on the fragment structures of lysine, the corresponding structures of specific product ions *m/z* 207 ($[K^*-84-H_2O+H]^+$), *m/z* 179 ($[K^*-84-HCO_2H+H]^+$), and *m/z* 162 ($[K^*-84-HCO_2H-NH_3+H]^+$) in the MS^3 spectra were proposed (Fig. 3.9).

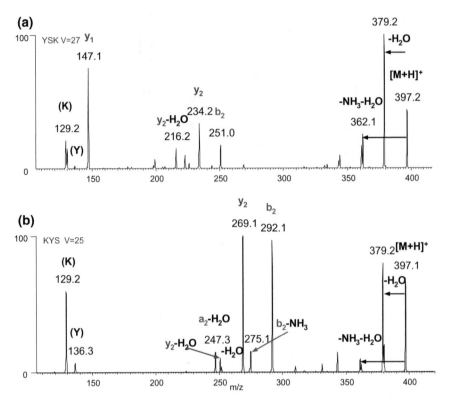

Fig. 3.7 MS/MS spectra of the two tri-peptides: **a** YSK and **b** KYS

3.6.3 MS³ *Spectra of [M-84+H]⁺ from Glucosylated Tri-Peptides*

The ions of $[M-84+H]^+$ from glycated tri-peptides with different lysine positions (KYS at N-terminal and YSK at C-terminal) were also studied by MS^3 experiments. MS^3 spectra of the two modified tri-peptides KYS and YSK are shown in Fig. 3.10. In MS^3 spectra of $[M-84+H]^+$ of glycated KYS, the modified product ions of b^*_2 and $b^*_2-NH_3$ (m/z 370.1 and m/z 353.1, respectively), a^*_2 and $a^*_2-NH_3$ (m/z 342.3 and m/z 325.3, respectively) were observed, while the unmodified y_2-ion (m/z 269.2) was detected with low relative intensity. For the glycated YSK, the MS^3 spectra of $[M-84+H]^+$ show the modified product ions of y^*_1 and y^*_2 (m/z 225.3 and m/z 312.3, respectively) and unmodified b_2-ion (m/z 251.1). Based on the assignment of these unmodified and modified y-, b-, and a-series ions in MS^3 spectra, peptide sequence information could be obtained [21]. From the modified and unmodified products of tri-peptides, MS^3 of $[M-84+H]^+$ shows the glycation site of peptides clearly. For example, modified y^*_1 and y^*_2 ions and unmodified b_2-ion of YSK show clearly that the modification site was at C-terminal. Besides,

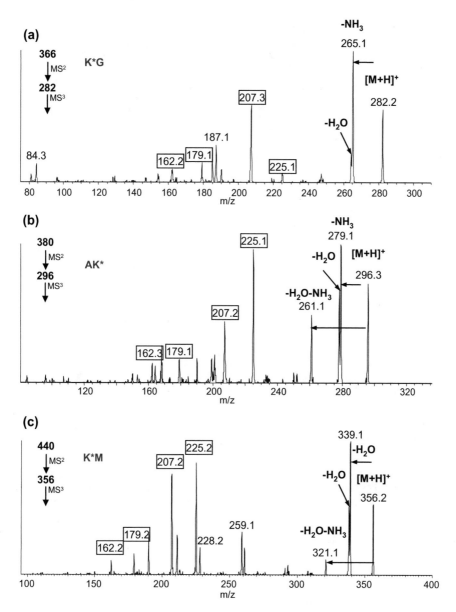

Fig. 3.8 MS³ spectra of [M-84+H]⁺ from the glucosylated di-peptides of **a** K*G, **b** AK*, and **c** K*M

the classical modified and unmodified N-terminal and C-terminal product ions, other typical product ions of ARPs from lysine with sugars were also used to verify lysine as the glycated site, e.g., ions of m/z 225 and m/z 207 in MS³ spectra, which were glycated lysine related product ions (Fig. 3.10).

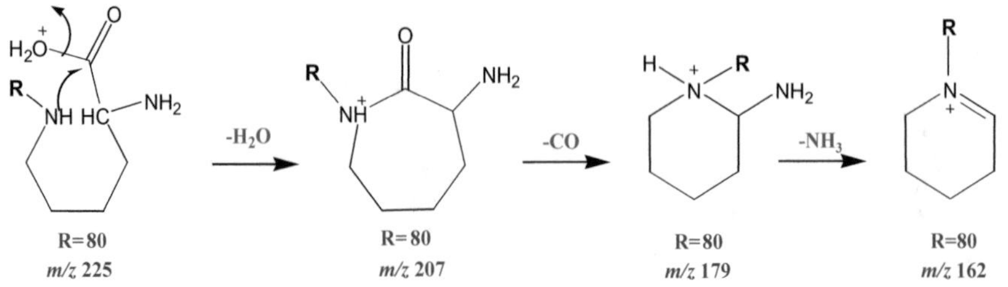

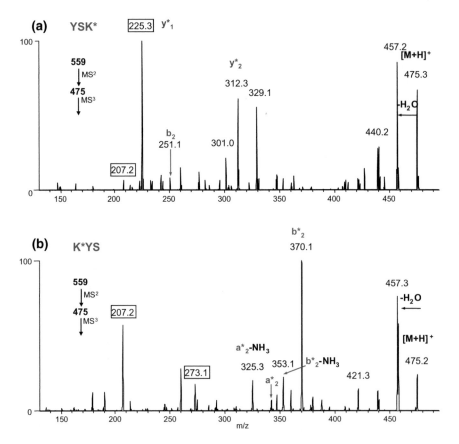

Fig. 3.9 Proposed fragmentation of glycated lysine and lysine-containing peptides via the cyclical mechanism

Fig. 3.10 MS^3 spectra of [M-84+H]$^+$ from the glucosylated tri-peptides of **a** YSK* and **b** K*YS

To provide more evidence of the use of MS/MS techniques for glycation site identification, MS^3 experiments of [M-84+H]$^+$ from three other tri-peptides (TKY, KYT, and KTY) were carried out.

Similarly, MS^3 spectra of $[M-84+H]^+$ of these three glycated peptides not only provided the sequence information, but also displayed the modification sites of lysine residue. The classical modified and unmodified N-terminal and C-terminal product ions are shown in Fig. 3.11. For glycated TKY, the modified product ion of b^*_2 and b^*_2-NH_3 (m/z 308.1 and 291.1, respectively), y^*_2 (m/z 388) and a^*_2-H_2O (m/z 264.0) were observed (Fig. 3.11a). For the glycated KYT and KTY which had the same lysine position at the N-terminal, MS^3 spectra of $[M-84+H]^+$ show the same unmodified y_2-ion (m/z 361.2) but different modified a^*_2- and b^*_2-ions. In MS^3 spectra of $[M-84+H]^+$ of glycated KYT, modified b^*_2 and b^*_2-NH_3 were at m/z 370.3 and m/z 353.3; a^*_2 and a^*_2-NH_3 were at m/z 342.1 and m/z 325.1, while in glycated KTY, b^*_2 and b^*_2-NH_3 of K^*TY were at m/z 308.1 and 291.0; a^*_2-H_2O was at m/z 264.1. Based on the modified N-terminal and C-terminal product ions, the glycated tri-peptides sequences could be identified. Moreover, the glycation sites of the three modified tri-peptides could also be verified with the typical product ions of glycated lysine residue related, like ions of m/z 207, m/z 179, and m/z 162 in MS^3 spectra, formed between immonium ion of lysine and oxonium ion of sugar skeleton, respectively (Fig. 3.11).

3.7 Fragmentation of Lysine-Containing Tetra-Peptides

Based on the MS data of the di-peptides and tri-peptides, two lysine-containing tetra-peptides (KGGK, KWGK, and KGWK) were synthesized to expand the application of MS techniques in glycation site identification. The glucosylated tetra-peptides displayed similar fragmentation behavior with the studied di-peptides and tri-peptides. At relatively low collision energy, the fragments ions of modified tetra-peptides were firstly produced from cleavage of sugar moiety with the loss of water molecules, while the product ions correlated to glucosylated lysine were observed at low mass range. MS/MS spectra of ARPs of tetra-peptides reacting with glucose also provided the sequence information with part of y-series ions (y_2-ion of m/z 330.3 and y_3-ion of m/z 390.2), besides the typical product ions from sugar moiety (Fig. 3.12). For the peptide moiety of the modified tetra-peptides, the specific product ions (m/z 291, m/z 273, m/z 255, and m/z 207) of the fragments derived from lysine residues were observed. The data suggest that the glycation site of the tetra-peptides should be the lysine residue although the modification site cannot be identified.

Since the MS/MS data only show few N-terminal and C-terminal product ions, it was difficult to perform peptide sequence mapping of the tetra-peptides. To obtain more sequence information, further MS^3 experiments of singly- and doubly-charged $[M-84+H]^+$ were conducted (Fig. 3.13). Spectra provide the fragmentation ions to identify the peptide sequence and glycation site clearly (Fig. 3.13). Both in singly- and doubly-charged MS^3 of $[M-84+H]^+$, unmodified y_3-ion (m/z 390) and modified a_2^*, b_2^*, b_3^*-serial ions such as m/z 365, m/z 393, and m/z 450 were observed; whereas, in the doubly charged MS^3 spectra of $[M-84+H]^+$,

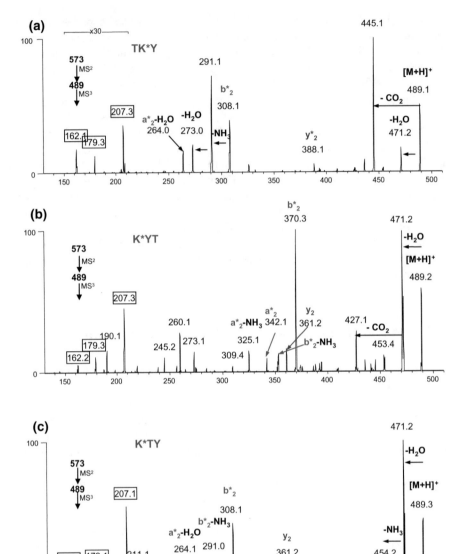

Fig. 3.11 MS^3 spectra of [M-84+H]$^+$ from the glucosylated tri-peptides of **a** TK*Y, **b** K*YT, and **c** K*TY

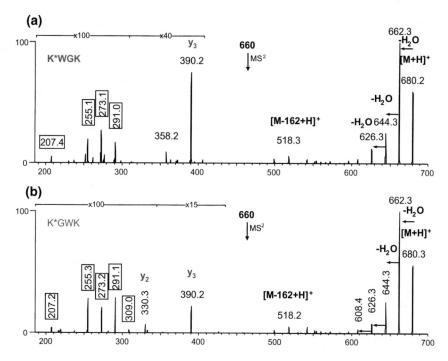

Fig. 3.12 MS/MS spectra of *m/z* 660 of glucosylated **a** K*WGK **a** and **b** K*GWK

only one strong y_1-ion was found at *m/z* 147.1 (Fig. 3.13b, d). With these serial product ions of the tetra-peptides, the sequence of tetra-peptides was identified which could provide useful information to verify the modification sites. For the studied tetra-peptides (KGGK, KWGK and KGWK), there were two available lysine to react with glucose. Based on the assignment of the serial product ions of glucosylated fragments in MS^3 spectra such as b_3*-ions (*m/z* 450) and b_3*+H$_2$O ions (*m/z* 468), the glycation sites of the two tetra-peptides were verified as the N-terminal lysine.

By comparison of singly- and doubly-charged MS^3 spectra of the tetra-peptides (Fig. 3.13), doubly charged spectra obviously provided more product ions with higher relative intensity at lower collision energy. Moreover, the additional fragment ions such as ions at *m/z* 129 and *m/z* 147 could be observed at low mass range in the doubly charged MS/MS spectra, due to expanding mass range by decreasing the effect of low mass cutoff in ion trap MS. The advantages of doubly or multiply charges in MS/MS experiments are crucial to analysis of enzymatically treated polypeptide and proteins, such as the characterization of the digested glycated proteins by LC-MS/MS.

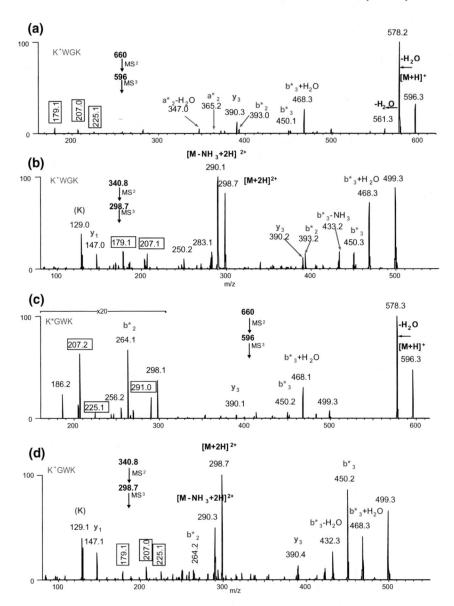

Fig. 3.13 MS^3 spectra of [M-84+H]$^+$ of **K*WGK a** singly charge and **b** doubly charge; MS^3 spectra of [M-84+H]$^+$ of **K*GWK c** singly charge and **d** doubly charge

3.8 Summary

The MS results show that reactants and reaction conditions affect the rate of Maillard reaction. The glycation rate was accelerated by increase in temperature, but the position of lysine had no obvious effect. MS/MS experiments show that fragment ions of ARPs exhibited characteristic fragmentation patterns.

MS/MS spectral data suggest that lysine residues were the preferential target of modification in the Maillard reaction, but it was difficult to acquire enough sequence information to characterize the peptide, due to non-specific cleavage properties of ion trap MS. MS^3 experiments of specific peak of [M-84+H]$^+$ could confirm the glycation site with direct fragmentation information.

References

1. Maillard LC (1912) Reaction of amino acids on sugars: formation of melanoidins by a systematic way. Compt Rend Acad Sci 154:66–68
2. Hodge JE (1955) The amadori rearrangement. Adv Carbohydr Chem 10:169–205
3. Adrian J, Frangne R (1973) Maillard reaction 8. Role of premelanoidins on nitrogen digestibility in-vivo and on proteolyse in-vitro. Ann Nutr Alimentation 27(3):111–123
4. Tagami U, Akashi S, Mizukoshi T et al (2000) Structural studies of the Maillard reaction products of a protein using ion trap mass spectrometry. J Mass Spectrom 35(2):131–138
5. Friedman M (1996) Food browning and its prevention: an overview. J Agric Food Chem 44 (3):631–653
6. Vinale F, Monti SM, Panunzi B et al (1999) Convenient synthesis of lactuloselysine and its use for LC-MS analysis in milk-like model systems. J Agric Food Chem 47:4700–4706
7. Sanz ML, Corzo-Martinez M, Rastall RA et al (2007) Characterization and in vitro digestibility of bovine beta-lactoglobulin glycated with galactooligosaccharides. J Agric Food Chem 55(19):7916–7925
8. Mennella C, Visciano M, Napolitano A et al (2006) Glycation of lysine-containing dipeptides. J Pept Sci 12(4):291–296
9. Jeric I, Versluis C, Horvat S et al (2002) Tracing glycoprotein structures: electron ionization tandem mass spectrometric analysis of sugar-peptide adducts. J Mass Spectrom 37(8): 803–811
10. Frolov A, Hoffmann P, Hoffmann R (2006) Fragmentation behavior of glycated peptides derived from D-glucose, D-fructose and D-ribose in tandem mass spectrometry. J Mass Spectrom 41(11):1459–1469
11. Iberg N, Fluckiger R (1986) Nonenzymatic glycosylation (glycation) of proteins—the principal sites of invitro glycation of rnase A. Experientia 42(6):680
12. Venkatraman S, Chan CM (1986) A novel method for cross-linking polyetherketones. Abstr Pap Am Chem Soc 191:118-POLY
13. Nakanishi T, Iguchi K, Shimizu A (2003) Method for hemoglobin A(1c) measurement based on peptide analysis by electrospray ionization mass spectrometry with deuterium-labeled synthetic peptides as internal standards. Clin Chem 49(5):829–831
14. Ahmed N, Thornalley PJ (2003) Quantitative screening of protein biomarkers of early glycation, advanced glycation, oxidation and nitrosation in cellular and extracellular proteins by tandem mass spectrometry multiple reaction monitoring. Biochem Soc Trans 31:1417–1422

15. Castro-Perez J, Plumb R, Liang L et al (2005) A high-throughput liquid chromatography/ tandem mass spectrometry method for screening glutathione conjugates using exact mass neutral loss acquisition. Rapid Commun Mass Spectrom 19(6):798–804
16. Scholz K, Dekant W, Volkel W et al (2005) Rapid detection and identification of N-acetyl-L-cysteine thioethers using constant neutral loss and theoretical multiple reaction monitoring combined with enhanced product-ion scans on a linear ion trap mass spectrometer. J Am Soc Mass Spectrom 16(12):1976–1984
17. Horvat S, Jakas A (2004) Peptide and amino acid glycation: new insights into the Maillard reaction. J Pept Sci 10(3):119–137
18. Jakas A, Katic A, Bionda N et al (2008) Glycation of a lysine-containing tetrapeptide by D-glucose and D-fructose—influence of different reaction conditions on the formation of Amadori/Heyns products. Carbohyd Res 343(14):2475–2480
19. Saraiva MA, Borges CM, Florencio MH (2006) Reactions of a modified lysine with aldehydic and diketonic dicarbonyl compounds: an electrospray mass spectrometry structure/activity study. J Mass Spectrom 41(2):216–228
20. Finot PA, Mauron J (1969) Lysine blockade via maillards reaction. I. Synthesis of N-(1-desoxy-d-fructos-1-yl)-l-lysine and n-(1-desoxy-d-lactulos-1-yl)-l-lysine. Helv Chim Acta 52(6):1488–1495
21. Ruan ED, Wang H, Ruan Y et al (2013) Study of fragmentation behavior of amadori rearrangement products in lysine-containing peptide model by tandem mass spectrometry. Eur J Mass Spectrom 19(4):295–303
22. Fogliano V, Monti SM, Visconti A et al (1998) Identification of a beta-lactoglobulin lactosylation site. Biochim Biophys Acta-Protein Struct Mol Enzymol 1388(2):295–304
23. Gadgil HS, Bondarenko PV, Treuheit MJ et al (2007) Screening and sequencing of glycated proteins by neutral loss scan LC/MS/MS method. Anal Chem 79(15):5991–5999
24. Li C, Wang H, Zhang Y et al (2014) Characteristics of early maillard reaction products by electrospray ionization mass spectrometry. Asian J Chem 26(21):7452–7456
25. Zhang Y, Ruan ED, Wang H et al (2014) A fundamental study of amadori rearrangement products in reducing sugar-amino acid model system by electrospray ionization mass spectrometry and computation. Asian J Chem 26(10):2914–2944
26. Yeboah FK, Yaylayan VA (2001) Analysis of glycated proteins by mass spectrometric techniques: qualitative and quantitative aspects. Nahrung-Food 45(3):164–171
27. Vinale F, Monti SM, Panunzi B et al (1999) Convenient synthesis of lactuloselysine and its use for LC-MS analysis in milk-like model systems. J Agric Food Chem 47(11):4700–4706
28. Morgan F, Bouhallab S, Molle D et al (1998) Lactolation of beta-lactoglobulin monitored by electrospray ionisation mass spectrometry. Int Dairy J 8(2):95–98
29. Monti SM, Ritieni A, Graziani G et al (1999) LC/MS analysis and antioxidative efficiency of Maillard reaction products from a lactose-lysine model system. J Agric Food Chem 47 (4):1506–1513
30. Fenaille F, Morgan F, Parisod V et al (2004) Solid-state glycation of beta-lactoglobulin by lactose and galactose: localization of the modified amino acids using mass spectrometric techniques. J Mass Spectrom 39(1):16–28
31. French SJ, Harper WJ, Kleinholz NM et al (2002) Maillard reaction induced lactose attachment to bovine beta-lactoglobulin: Electrospray ionization and matrix-assisted laser desorption/ionization examination. J Agric Food Chem 50(4):820–823
32. Huddleston MJ, Bean MF, Carr SA (1993) Collisional fragmentation of glycopeptides by electrospray ionization Lc Ms and Lc Ms Ms—methods for selective detection of glycopeptides in protein digests. Anal Chem 65(7):877–884

Chapter 4
Determination of the Maillard Reaction Sites and Properties' Effects of Lysozyme

4.1 Introduction

Foods are routinely subjected to various heat treatments during processing such as steaming, tempering, and roasting [1, 2]. These thermal treatments often lead to substantial denaturation or unfolding of proteins which influence protein functionality [3]. It had been established that functional properties of proteins can be improved by covalent bonding with sugars through the Maillard reaction, a naturally occurring reaction during food storage and/or cooking [4–8]. The Maillard reaction could be accelerated by high temperatures without extraneous chemicals, and it is probably the most promising approach to improve protein properties for food applications. Effects of protein–reducing sugar conjugation via the Maillard reaction on functional properties had been widely studied using model systems [9–15]. After modification with sugars, such as glucose, galactose, and lactose, some food proteins showed better emulsifying properties and the heat stability and solubility were also improved [4, 16]. Influence of glycation on thermal properties of proteins had also been studied [13, 17–21]. Differential scanning calorimetry (DSC) is a suitable technique for studying the thermal properties of proteins [22–26]. Denaturation temperature (T_d), width at half-peak height ($\Delta T_{1/2}$), and enthalpy change (ΔH) can be determined from the thermograms. T_d value is a measure of thermal stability of proteins. ΔH, measured as the area under the endothermic peak, represents the proportion of undenatured protein [5, 27]. The $\Delta T_{1/2}$ value is an index to show the cooperativity of the denaturation process. Normally, a smaller $\Delta T_{1/2}$ value shows the higher cooperativity [28]. Thermal denaturation of proteins involves conformational changes due to the disruption of chemical forces that maintain the structural integrity of the molecules [22, 23].

The study of the biological properties of Maillard reaction usually demonstrates only the formation of reaction product and lacks detailed information about the relationship of the glycation site structure and the corresponding biological activity [29, 30]. It is therefore desirable to develop analytical methods for the detection of

© The Author(s), under exclusive license to Springer Nature Switzerland AG 2018
D. Ruan et al., *The Maillard Reaction in Food Chemistry*, Chemistry of Foods
https://doi.org/10.1007/978-3-030-04777-1_4

protein-bound Maillard products in order to understand the structure–function relationship. Although the Maillard reaction has been studied in vitro or in vivo for many years, the detection of protein glycation products and the identification of glycation sites have always been a challenging area of investigation. The previous analyses of protein glycation mainly focused on a mass increase in intact proteins, either isolated from sources in vivo or produced upon incubation of carrier proteins with reducing sugars, and the primary glycated sites of modified proteins have not yet been clarified and analyzed [31–34]. Recently, ESI-MS and MALDI-TOF had been used to characterize posttranslational modifications of glycated proteins in the Maillard reaction [31, 32, 35–40].

In the present study, the glycation process in a lysozyme–glucose model was monitored under temperature and time control by MS-based techniques. MALDI-TOF-MS and ESI-MS can provide precise and accurate information of the reaction level and the progressive binding of glucose with lysozyme during incubation. The effects of glucosylation on emulsifying and thermal properties of lysozyme were also evaluated, and the extent of glucosylation was optimized based on the MS data. Furthermore, MALDI-TOF/TOF-MS and nano-LC-QqTOF-MS were used to analyze the specific products by mass shift, to demonstrate structural changes and to confirm the glycation sites. Site-specific peptide mapping of the enzymatically digested unmodified and modified lysozyme was conducted. Owing to different collision energy levels, TOF/TOF-MS provides more fragmentation ions of glycated peptide by high collision energy, whereas QqTOF-MS shows more cleavage of neutral losses of the sugar moiety. By combining of MALDI-TOF/TOF-MS and nano-LC-QqTOF-MS, four glycation sites of the Maillard products were determined and identified successfully.

4.2 Sample Preparation

LYZ (10 g) and D-glucose (1.0 g) were dissolved in water (100 mL) to give a protein-to-sugar molar ratio of 1:6. Aliquots of the protein–sugar mixture solution (5 mL) were placed in 15 mL plastic test tubes covered with perforated aluminum foil and incubated at 50 and 70 °C in an oven. The samples were incubated for 0–40 days in sealed 50-mL test tubes. Samples were presented as LYZ-G^{50}-x and LYZ-G^{70}-x, respectively (G: glucose; 50 and 70: temperature in °C; x: incubation time in days). After incubation, the reaction mixture was dialyzed with ultrapure water at 4 °C. The water was changed frequently for 2 days, with gentle stirring during the process. After dialysis, the samples were freeze-dried and then stored at −20 °C until they were analyzed.

LYZ (10.0 g) and D-glucose (1.0 g) were dissolved in water (100 Ml) to give an approximate protein-to-sugar molar ratio of 1:6. Aliquots of the protein–sugar mixture (5 mL) were frozen and lyophilized. Aliquots of the dried mixture were sealed in plastic test tubes and incubated at 50 °C for up to 24 h in an oven. Samples were presented as LYZ-G-x (x: incubation time in hours). After incubation, the samples were stored at −20 °C until they were analyzed.

4.3 MS

4.3.1 MALDI-TOF-MS

MALDI-TOF-MS spectra were acquired using an Applied Biosystems (Foster City, CA) 4800 Proteomic Analyzer operated in the positive MS mode at 2 kV. The peak intensity was determined using DataExplorer software (Applied Biosystems). α-CHCA and sinapinic acid were used as matrices for digested proteins and intact proteins, respectively. The MALDI matrix (10 mg/mL) was prepared by dissolving the recrystallized matrix in 50:50 (v/v) water/ACN incorporating 0.1% TFA. The instrument was equipped with a nitrogen laser (λ 337 nm) and a reflector. For MALDI-TOF-MS, laser-desorbed positive ions were analyzed after acceleration at 20 kV in the linear mode for the intact protein and at 19 kV in the reflector mode for the enzymatically digested protein, with one fixed laser intensity and 4500 laser shots per spectrum. Each spot was calibrated through external calibration, spotted in a separate sample spot. External calibration was performed using a standard peptide/protein mixture.

4.3.2 MALDI-TOF/TOF-MS

MALDI-TOF/TOF-MS was conducted using 2 kV collision energy with air as collision gas. MS/MS analysis was performed either manually or automatically with a laser intensity that was 40% higher than that used for MS analysis. MS/MS was performed until the S/N ratio was 30 for at least 20 peaks in the spectrum, with an upper limit of 15,000 shots per spectrum. The instrument was equipped with a nitrogen laser (λ 337 nm) and a reflector. Instrument settings for MS/MS were: ion source 1 potential, 8.00 kV; ion source 2 potential, 7.20 kV; reflectron 1 potential, 29.50 kV; reflectron 2 potential, 13.75 kV; LIFT 1 voltage, 19.00 kV; LIFT 2 voltage, 3.00 kV. The ion selector resolution was set at 0.5% of the mass of the precursor ion. The peak intensity was determined using DataExplorer software (Applied Biosystems).

4.3.3 Nano-LC-QqTOF-MS

In-house packed C18 reverse-phase (RP) fused silica capillary (80 cm × 150 μm i.d. (inner diameter); Polymicro Technologies, Phoenix, AZ) and trap columns were prepared under constant pressures of up to 9000 psi using ultrahigh-pressure 65D syringe pumps (UPLC; Isco, Lincoln, NE). The packing material was C18 particles (4 μm; 90   pore size; Phenomenex, Torrance, CA); the packed capillary was directly connected to the valves using PEEK tubing (308 μm i.d.; Upchurch Scientific,

Oak Harbor, WA) with a stainless-steel mesh screen (2 μm pores; Valco). A 10–15-μm i.d. outlet tip, made in-house, was used as the emitter which was fixed via a stainless-steel connect for high voltage to the mass spectrometer. The mobile phases A (0.5% aqueous formic acid) and B [ACN/water (98:2, v/v) containing 0.5% formic acid] were connected to a stainless-steel mixer (ca. 1.5 mL) containing a magnetic stirrer prior to elution (UPLC) into the capillary columns under a constant pressure of 8000 psi.

Sample trapping was performed online for 20 min, followed by a 3.5-h RP gradient and then a 3.5-h NP gradient. The RP-LC eluent was directly electrosprayed into a QSTAR XL quadrupole/TOF hybrid mass spectrometer (Applied Biosystems, Framingham, MA) operated under information-dependent acquisition. The mass spectrometric parameters were set as follows: DP, 20.0 V; FP, 125.0 V; DP, 30 psi; CG, 3.0 psi; GS1, 0 psi; GS2, 0 psi; CUR, 20 psi; nano-spray voltage, 3000 V. Rolling collision energy and enhanced ion settings were used in the MS/MS experiment; the linear calibration curve of the collision energy was optimized through the direct infusion of bovine albumin tryptic digest. For the RP Nano-LC-QqTOF-MS experiments, MS/MS data were first converted into SEQUEST format and then searched against the SEQUEST algorithm using BioWorks 3.3 (ThermoQuest Corp, Waltham, MA). Two missed cleavages were set. The peptide and fragment ion tolerances were set at 100 ppm and 0.3 Da, respectively.

IDA was performed in the Analyst software without using the inclusion and exclusion list. IDA included a 1-s survey scan followed by three 1-s product ion scans of the three most intense peaks having charge states ranging from 2 to 5. The mass-to-charge ratios of the fragmented ions were placed on an exclusion list for 20 s to avoid redundant analysis. An intensity threshold of 50 counts was set to trigger the product ion scan.

MS/MS data were first converted into SEQUEST format and then searched against the SEQUEST algorithm using BioWorks 3.3 (ThermoQuest Corp, Waltham, MA). Two missed cleavages were set. The peptide and fragment ion tolerances were set at 100 ppm and 0.3 Da, respectively.

4.4   Emulsifying Properties

Emulsification activity index (EAI) and emulsion stability index (ESI*, showing difference with ESI of electrospray ionization) of LYZ and glucosylated LYZ were determined by the turbidimetric method [41]. Aliquots of 22.5 mg protein were weighed into plastic cups, and 4.5 mL phosphate buffer (pH 7.0) and 1.5 mL corn oil were added. The solution was magnetically stirred for 30 min at room temperature and then homogenized for 1 min with an Omni Mixer Homogenizer equipped with a micro-attachment at a speed of 8000 rpm. An aliquot (250 μL) of emulsion was pipetted from the bottom of the cup into 50-mL volumetric flask, and 0.1% SDS was added to fill to the mark. The absorbance of the diluted and stabilized emulsion was measured at 500 nm. Another aliquot (0.8 mL) of the emulsion was placed into a desiccator and reweighed after room temperature was reached. For emulsion stability determination, aliquots of emulsion were pipetted and diluted with 0.1% SDS at appropriate time intervals (about 2–3 min) until the absorbance decreased to below 75% the initial value. ESI[*] was defined as the time in minutes for A_{500} to decrease to 75% of the value at zero-time.

Calculation:

$$\Phi = [(C-A)-E(B-C)]/[(C-A)+(B-C) * 0.918]$$

$$\mathrm{EAI}\,(\mathrm{m}^2/\mathrm{g}) = [4.606 * \mathrm{A}500 * \text{dilution factor}]/[\Phi * \text{conc. of matter} * 10,000]$$

where Φ = oil volume fraction; EAI = emulsifying activity index (m^2/g); A = weight of aluminum dish; B = weight of aluminum dish + 0.8 mL emulsion; C = weight of aluminum dish + dried matter; and E = concentration of protein solution (g/mL).

Also, ESI* could be calculated as the following equation:

$$\mathrm{ESI*}\,(\mathrm{min}) = A_0/(A_0-A_{10}) * 10$$

where A_0 = absorbance value of zero-time at 500 nm; A_{10} = absorbance value of 10 min at 500 nm.

4.5 Differential Scanning Calorimetry (DSC)

Thermal properties of glucosylated LYZ under dry-heating were examined by TA 2920 modulated DSC thermal analyzer (TA Instruments, New Castle, DE, USA). Approximately 1 mg of protein was directly weighed into the polymer-coated aluminum pan, and 10 μL of different buffer solutions was added. The pan was hermetically sealed with a lid and equilibrated for 15 min before DSC analysis. The pans were heated at a rate of 10 °C/min from 25–140 °C. A sealed empty pan was used as a reference. Indium standards were used for temperature and energy calibration. Thermal transition characteristics including T_d, ΔH, and $\Delta T_{1/2}$ were computed from the thermograms by the Universal Analysis Program, version 2.5H (TA Instruments, New Castle, DE, USA). All experiments were conducted in duplicate, and results were reported as the means of these duplicates ± standard deviations.

Emulsifying property experiments and DSC experiments had three replications. To determine statistically significant differences between samples ($p < 0.05$), the data were subjected to analysis of variance and appropriate means separation was conducted using Duncan's multiple range test using a statistical software program (SPSS for Windows Version 7.0).

4.6 Deconvolution of ESI-MS Spectra

ESI-MS spectra could provide molecular weight from ions with multiple charges [37, 42]. The ESI-MS spectra of lysozyme were shown in Fig. 4.1.

The accurate molecular weight of the intact protein can be determined through deconvolution of the spectrum using the ProMass software of the Finnigan Xcalibur data system. The deconvoluted spectra allowed rapid visualization of the results, including an integrated sample plate viewer. The summary for all of the processed samples generates chromatograms automatically labeled with the charge states of the most abundant components in Fig. 4.2.

4.7 MS Analysis of Intact Lysozyme

Although SDS-PAGE was commonly used to separate non-fragmented glycated proteins, it was limited to the analysis of highly modified proteins that exhibited large mass increases for clear separation [6, 43–45]. In comparison with traditional SDS-PAGE analysis methods, MALDI-TOF mass spectrometric techniques could obtain more accurate molecular weight of native and modified proteins directly and quickly and monitor the molecular mass changes during the modification without further purification [43, 46, 47]. Figure 4.3 shows typical MALDI-TOF-MS spectra of wet-heated LYZ at 50 °C for various time intervals. The glycated LYZ exhibited distinct broadening and shifts of the peaks toward higher masses (ca. 15,000 Da). It was apparent that the protein glycoforms had a heterogeneous distribution.

The control sample, LYZ-G^{50}-0, exhibited a sharp and dominant mass peak at m/z 14.3 K in the MALDI-TOF-MS spectrum (Fig. 4.3a), while the spectrum of the LYZ-G^{50}-1 (Fig. 4.3b) was similar except that the peak width had increased slightly. After incubation for one week (LYZ-G^{50}-7; Fig. 4.3c), significant peak broadening appeared and the heterogeneous peaks shifted to higher masses with a

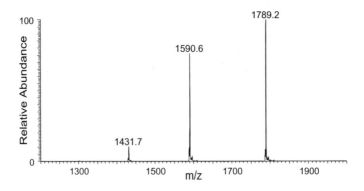

Fig. 4.1 ESI-MS spectrum of intact lysozyme

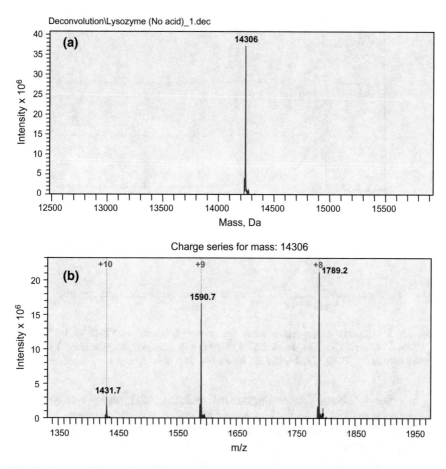

Fig. 4.2 Deconvolution of ESI-MS spectra of **a** monocharged protein and **b** multiply charged protein using ProMass software

decrease in the intensity of the signal at m/z 14.3 K. The results reveal that the extent of Maillard reaction increased with increasing incubation time. The disappearance of the original peak at m/z 14.3 K was accompanied by the appearance of a broad peak having a maximum m/z value at 15 K (Fig. 4.3e), demonstrating that lysozyme had reacted with the sugar.

MALDI-TOF-MS spectra of intact protein provide molecular weight directly and information about the extent of modification caused by multiple modification. Specific modification products could not be resolved, probably due to the fact that the mass increases were too small to be separated from the intact protein and the resolution of the technique in this mass range was not high enough to identify defined products, rendering it difficult to obtain the accurate glycation data. The insufficient resolution was due to the singly charge peak of the MALDI-TOF-MS.

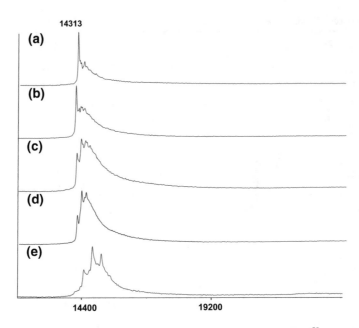

Fig. 4.3 MALDI-TOF-MS spectra of intact glucosylated lysozyme. **a** LYZ-G^{50}-0, **b** LYZ-G^{50}-1, **c** LYZ-G^{50}-7, **d** LYZ-G^{50}-14, and **e** LYZ-G^{50}-30 (LYZ: lysozyme; G: glucose; ^{50}: incubation temperature in 50 °C; 0, 1, 7, 14, and 30: incubation time in days)

In order to obtain further information, reaction mixtures of lysozyme with D-glucose were also submitted to nano-ESI-MS to monitor changes during the glycation process. Compared to MALDI-TOF-MS spectra with singly charge and broad peak, ESI-MS spectra show intact lysozyme and glycated lysozyme with their multiply charged ions. Figure 4.4 shows the ESI-MS mass spectrum of intact lysozyme ions featuring charges of +8, +9, and +10 and the progress of the Maillard reaction when lysozyme was incubated with D-glucose at 50 °C for 1, 7, 14, and 30 days. Through charge calculation, performed as described previously, the peaks at 1431.7, 1590.6, and 1789.2 in the spectrum of LYZ-G^{50}-0 corresponded to charges of +10, +9, and +8, respectively (Fig. 4.4a). One additional peak appeared in each peak cluster in the spectrum of LYZ-G^{50}-1 with the mass increase of 162 Da. The ESI-MS spectrum of LYZ-G^{50}-30 shows that four molecules of glucose were covalently attached to lysozyme, as revealed by the four new peaks in each of the three charged states (Fig. 4.4e). The results show a distribution of different glycoforms of lysozyme at different multiply charged states, and each peak in a cluster at a particular charged state represents a glycoform of lysozyme (ARPs) containing a specific number of sugar moieties. The ESI-MS spectra of intact proteins clearly show new peaks, and the molecular mass difference between adjacent molecular ion peaks within a cluster was determined to be 162 Da, equivalent to the mass of a covalently bound hexose moiety.

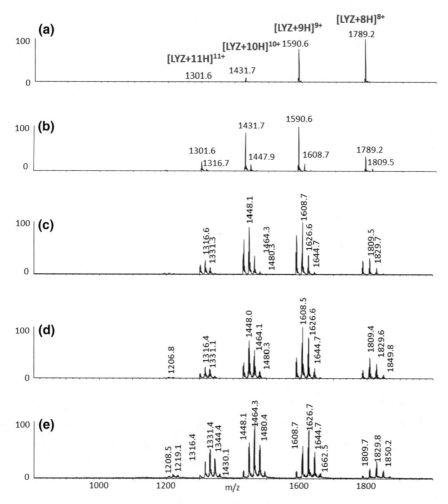

Fig. 4.4 ESI-MS spectrum of intact glucosylated lysozyme with multiple charges. **a** LYZ-G^{50}-0, **b** LYZ-G^{50}-1, **c** LYZ-G^{50}-7, **d** LYZ-G^{50}-14, and **e** LYZ-G^{50}-30 (LYZ: lysozyme; G: glucose; ^{50}: incubation temperature in 50 °C; 0, 1, 7, 14, and 30: incubation time of days)

The Maillard reaction rates of wet-heated (LYZ-G^{50}-x and LYZ-G^{70}-x) and dry-heated (LYZ-G-x) LYZ were determined from the average sugar loading values (SLVs) per molecule of lysozyme at different incubation times by ESI-MS. A plot of SLVs of lysozyme versus time of incubation is shown in Fig. 4.5. For wet-heating, it had been observed that the SLVs during the initial stages (0–14 days) of the Maillard reaction solution were influenced mainly by temperature, but the temperature effect diminished during the next stage (14–30 days). At 70 °C, however, the rate of sugar loading of lysozyme increased from 1.0 to 3.0 between the first and tenth days of incubation, as compared to a change from 0.8 to

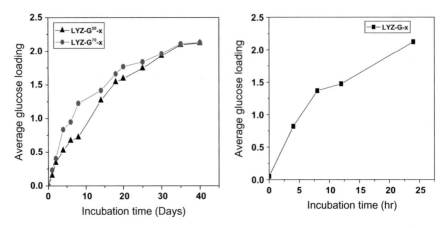

Fig. 4.5 Average sugar loading values of a lysozyme–glucose system during (Left) wet-heating and (Right) dry-heating. The percentage distribution of the different glycoforms of monomeric lysozyme in the incubation mixtures at different stages of incubation is presented in Fig. 4.6

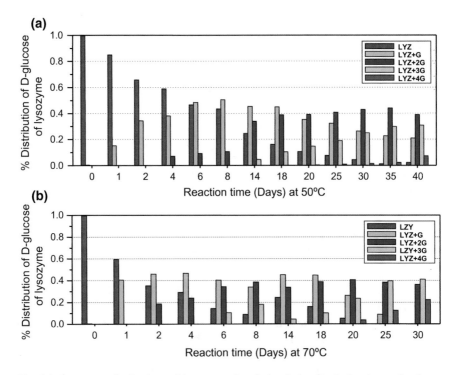

Fig. 4.6 Percentage distributions of lysozyme glycoforms during incubation in wet-heating at **a** 50 °C and **b** 70 °C

2.0 at 50 °C. The change of temperature did not have an apparent effect on the initial rate of sugar loading in the lysozyme–sugar system, although the overall sugar loading value of lysozyme at 70 °C was slightly high. The changes in SLV during dry-heating were similar to those during wet-heating, except that the SLV of the dry-heated samples increased much faster at 50 °C. Dry-heating for 4 h produced an SLV similar to that of the wet-heated sample for 10 days. It took 30 days to obtain an SLV of 2.0 through wet-heating at 50 or 70 °C, whereas through dry-heating, it was accomplished only within 24 h. The Maillard reaction proceeded at a rapid pace during the initial stages because the maximum number of lysine ε-amino groups was accessible for interaction with glucose. Upon increasing the incubation time, a saturation point was reached as a result of a decrease in either the protein's flexibility or the accessibility of the remaining amino groups. The extent of the Maillard reaction was also related to the molecular weight of the reducing sugar.

Panels A and B represent lysozyme with D-glucose samples incubated at 50 and 70 °C, respectively. A comparison of panel A and B reveals important differences between the glycation of lysozyme at different temperatures. With similar SLVs for LYZ-G^{50}-30 (1.93) and LYZ-G^{70}-30 (1.96), the five different glycoforms of LYZ-G^{50}-30 contained 0–4 equivalents of glucose, whereas three glycoforms of the LYZ-G^{70}-30 contained 2–4 equivalents of glucose. Di-glycosylated lysozyme was the dominant glycoform in LYZ-G^{50}-30, whereas the dominant glycoform of LYZ-G^{70}-30 was tri-glycosylated. The results show that the Maillard reaction products with similar SLVs might comprise of different glycoforms of protein.

4.8 Effects of Glucosylation on Lysozyme

4.8.1 Effect on Emulsifying Properties

Emulsion activity index (EAI) of a protein is a measure of the capacity of the protein to remain at oil–water interface after emulsion formation and is highly affected by the balance of hydrophobic and hydrophilic properties of the protein–sugar conjugates [15, 48]. On the other hand, emulsion stability index (ESI) is an evaluation of protein's capacity to remain at oil–water interface during storage or heating [49]. The emulsifying properties of glucosylated lysozyme samples (LYZ-G-x, x = 0, 4, 8, 12, and 24) are summarized in Table 4.1.

Glucosylation led to a progressive increase in EAI at the beginning of the reaction, and a maximum EAI value was obtained after 8 h by dry-heating. The EAI values of LYZ-G-4 (52.1 m^2/g) and LYZ-G-8 (109.3 m^2/g) were significantly ($p < 0.05$) higher than that of the unheated control (41.5 m^2/g). EAI was decreased upon further heating. LYZ-G-12 still had an EAI (64.3 m^2/g) significantly higher than that of LYZ-G-0. A dramatic decrease in EAI value was observed when the incubation time was extended to 24 h. The results showed that incubation time had

Table 4.1 Emulsifying properties of native and glucosylated lysozyme produced by dry-heating at 50 °C

Heating time (days)	EAI (m²/g)	ESI (min)
0 (native)	41.5 ± 2.128^b	19.7 ± 0.290^a
4	52.1 ± 1.881^c	25.9 ± 0.869^b
8	109.3 ± 1.896^d	27.7 ± 0.128^c
12	64.3 ± 1.728^c	33.7 ± 0.349^d
24	20.3 ± 0.737^a	40.7 ± 0.571^e

Data are means $\pm$S.D. of 3 replicates. Means within the same column with different letters (a, b, c, d) are significantly different ($p < 0.05$)

a pronounced influence on emulsification activity of lysozyme, with a positive effect upon mild glucosylation, and a negative effect upon extensive glucosylation. Consistent with the MS data, the SLVs of LYZ-G-x ($x = 4$, 8 and 12) were in the range of 0.8–1.5 in the glucosylated lysozyme mixture. However, the SLV of LYZ-G-24 was over 2.2 and di-glucosylated lysozyme was the dominant glyco-form. Hence, maximum EAI value was obtained when glucosylated lysozyme was mainly constituted of monoglycoform.

Glucosylation by dry-heating significantly ($p < 0.05$) increased the ESI value of lysozyme (Table 4.1). A gradual and significant increase in ESI was observed with increasing reaction time. It was assumed that the N-terminal and/or the other three ε-amino groups were glucosylated preferentially, and the structure of glucosylated lysozyme could have more amphiphilic characteristic. This characteristic was important for the stability of emulsion against coalescence of droplets [50, 51].

Several previous studies also showed that glucosylation improved both the emulsifying activity and emulsion stability of native lysozyme, because lysozyme conjugated with sugars could have enhanced emulsifying properties [52–54]. When compared with other glucosylated proteins prepared by controlled dry-heating [55], the present data showed similar improvement in emulsifying properties. However, excessive glucosylation could lead to market decrease in EAI. It had been reported that maximum improvement in emulsifying properties was obtained at 46.7% of available lysine in lysozyme [56]. The decrease in EAI after extended glucosylation might be caused by excessive sugar bonding [57, 58]. During emulsification, the hydrophobic residues of proteins were anchored in the oil droplets, and the attached sugars attract water molecules around the oil droplets, accelerating the formation of a steric stable layer around the emulsion and inhibiting the coalescence of oil droplets, resulting in the increases in EAI and ESI. However, when the proportion of saccharides in the conjugate reaches a maximum level, the availability of the protein to adsorb at the oil–water interface was decreased, leading to a decrease in EAI [59–61]. Reversible depletion and flocculation caused by saccharides in the emulsion phase also led to the decrease in emulsion stability [62–64]. This influence could also be used to explain the size of reacted molecules which affects protein propagation at the oil–water interface [61, 65]. Therefore, the use of mono- and disaccharides and mild reaction condition of glucosylation could enhance emulsifying properties [66–69].

4.8.2 Effect on Thermal Properties

Figure 4.7 shows the DSC thermograms of glucosylated lysozyme samples incubated for 0, 4, 8, 12, and 24 h at 50 °C, and a single peak was observed in all the thermograms. Table 4.2 gives the thermal characteristics of native and glucosylated lysozyme. Mild glucosylation significantly ($p < 0.05$) increased the denaturation enthalpy (ΔH) of LYZ-G-4 (16.7 J/g) and LYZ-G-8 (16.5 J/g) when compared with the control (15.1 J/g), whereas extended reaction time significantly ($p < 0.05$) decreased ΔH. Glucosylation led to a slight increase in denaturation temperature (T_d) of LYZ-G-x ($x = 4$, 8 and 12). The width at half-peak height ($\Delta T_{1/2}$) value was increased by glucosylation, and the increase was significant ($p < 0.05$) at 12 and 24 h of incubation. According to the research work of Privalov [28], most proteins had insufficient hydrophobic interactions to maintain a "thermal core," preventing T_d to go above 110 °C, and this might be universal for all compact globular proteins.

The T_d values of glycosylated lysozyme samples were observed at a narrow range from 80.6 to 82.2 °C in the present study, higher than that (69–76 °C) measured by other researchers [17, 70, 71]. This could be due to the measurement at a faster heating rate of 10 °C/min instead of 1 °C/min. Heating rate had been shown to affect T_d measurements previously [72]. Furthermore, differences in purification steps could also lead to protein preparations with different DSC characteristics [73, 74].

Increases in thermal stability (as indicated by increases in T_d) in glucosylated lysozyme might be due to protein rearrangement to form a relative compact conformation, or association between the protein and sugar molecules to a complex

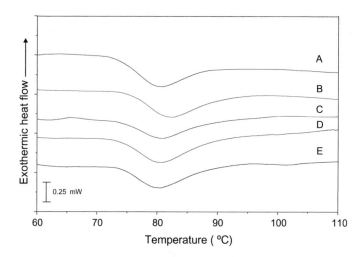

Fig. 4.7 DSC thermograms of a lysozyme–sugar system dry-heated at 50 °C. **a** LYZ-G-0, **b** LYZ-G-4, **c** LYZ-G-8, **d** LYZ-G-12, and **e** LYZ-G-24. LYZ: lysozyme; G: glucose; 0, 4, 8, 12, and 24: incubation time in days

Table 4.2 Thermal transition characteristics of native and glucosylated lysozyme produced by dry-heating at 50 °C

Heating time (days)	ΔH (J/g)	T_d (°C)	$\Delta T_{1/2}$ (°C)
0	15.1 ± 0.220^c	80.6 ± 0.346^a	8.3 ± 0.247^a
4	16.7 ± 0.898^d	82.2 ± 0.361^b	$9.0 \pm 0.050^{a,b}$
8	16.5 ± 0.495^d	81.2 ± 0.170^a	$9.0 \pm 0.424^{a,b}$
12	12.4 ± 0.431^b	80.8 ± 0.375^a	9.4 ± 0.254^b
24	11.0 ± 0.401^a	80.6 ± 0.184^a	9.3 ± 0.127^b

Data are means ±S.D. of 3 replicates. Means within the same column followed by different letters (a, b, c, d) are significantly different ($p < 0.05$)
ΔH Denaturation enthalpy, T_d denaturation temperature, $\Delta T_{1/2}$ width at half-peak height

aggregate structure with higher thermal stability. Unfolding of native protein also could cause the exposure of buried polar groups, enhancing protein–protein interactions by hydrophobic association [75, 76]. Subsequent heating of the modified proteins would require the rupture of more hydrophobic groups than the native protein, resulting in higher T_d [17, 18, 77]. Increases in T_d and decreases in ΔH in glucosylated lysozyme suggested that although the reaction was carried out at a mild temperature of 50 °C, much lower than the T_d of lysozyme (80 °C), protein conformation was perturbed to certain extent. The hydrophobic cores buried in the interior might become partially exposed, and the attached glucose molecules could also increase the stability of lysozyme within certain degree of glucosylation [78, 79].

As a result, the partially unfolded lysozyme could refold to form a more stable structure (aggregates) with higher T_d and ΔH. Significant decreases in ΔH in LYZ-G-12 and LYZ-G-24 indicated that protein denaturation and aggregation could occur during extended glucosylation. Moreover, breakup of hydrophobic interactions and protein aggregation were exothermic reactions, which could lower the net endothermic contribution causing a decrease in ΔH [80]. Protein denaturation might also lead to increases in $\Delta T_{1/2}$ value since a partially unfolded protein could denature in a less cooperative manner in the process of the Maillard reaction [28].

4.9 Characterization of Glycation of Lysozyme

4.9.1 MALDI-TOF-MS

MALDI-TOF-MS analysis of intact proteins is not sufficient to determine low molecular weight modifications (e.g., 162 Da) on heterogeneously glucosylated proteins (e.g., 14.3 kDa of lysozyme). Methods to identify these advanced glycation products using MALDI-TOF-MS peptide mapping were developed recently, and several advanced glycation end products have been identified and could

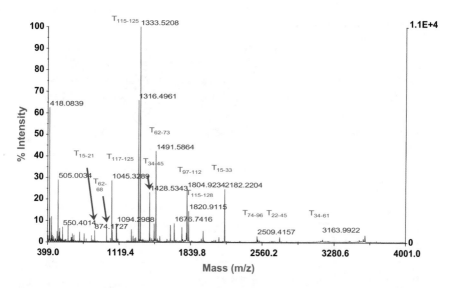

Fig. 4.8 MALDI-TOF-MS spectra of trypsin digested lysozyme (the spectra were converted into peak list format and then searched against the Mascot online (http://www.matrixscience.com)

therefore be used for structure assignments [47]. Therefore, digestion of proteins is necessary to obtain defined fragments of lower molecular range (580–3500 Da), which could result in a fine resolution for the relative mass increases due to modification on a molecular level (Fig. 4.8). Compared to theoretical digestion of lysozyme in the protein database [81, 82], it is possible to accomplish the assignments of the corresponding glycated peptide fragment according to the specific mass differences. Figure 4.9 shows the Mascot search results.

From the protein data bank [81, 82], the amino acid sequence and composition of lysozyme was obtained. By theoretically digesting lysozyme with trypsin, we were able to link an observed peak to a defined peptide fragment or amino acid sequence. The peak list was submitted to the Mascot for peptide mapping (coverage 55%). The purpose of this investigation is to establish MALDI-TOF-MS peptide mapping for the preliminary determination of tryptic-digested native protein and glycation on modified proteins. By comparing information from the database [81, 82] to the well-described chemical mechanisms for the early-stage Maillard reaction, the combination of Δm measured and the assigned modification structure seems very plausible.

Figure 4.10 shows the expanded spectra of glucosylated lysozyme in the ranges m/z 600–780 and 1760–2000, respectively. Prominent pairs of peptides were present at m/z 606.35/768.38 and m/z 1804.92/1966.99, differing by 162 Da and signifying the pairs of glycated lysozyme fragment.

Comparing the two panels, two additional peaks could be observed—the mass spectrum of glycated lysozyme exhibited two areas where new peaks appeared, representing the equivalent spectrum of native lysozyme. Figure 4.10a shows a

Sequence Coverage: **55%**

Matched peptides shown in **black**

1	11	21
KVFGR CELAA	AMK **RHGLDNY**	**R** GYSLGNWVC
31	41	51
AAK **FESNFNT**	**QATNRNTDGS**	**TDYGILQINS**
61	71	81
RWWCNDGR TP	GSRNLCNIPC	SALLSSDITA
91	101	111
SVNCA **KKIVS**	**DGNGMNAWVA**	**WR** NRC K **GTDV**
121		
QAWIR GCRL		

Start – End	Observed	Mr(expt)	Mr(calc)	Delta	Miss	Sequence
1-5	606.3447	606.3422	606.3401	0.0021	0	-.KVFGR.C
14 - 21	1030.5253	1029.5180	1029.5104	0.0076	1	K.RHGLDNYR.G
15 - 21	874.4084	873.4011	873.4093	-0.0082	0	R.HGLDNYR.G
34 - 45	1428.6831	1427.6759	1427.6429	0.0330	0	K.FESNFNTQATNR.N
46 - 61	1753.8837	1752.8764	1752.8278	0.0486	0	R.NTDGSTDYGILQINSR.W
62 - 68	993.4053	992.3980	992.3923	0.0057	0	R.WWCNDGR.T
97 - 112	1804.7510	1804.7656	1804.7592	0.0064	1	K.KIVSDGNGMNAWVAWR.N
98 - 112	1675.8346	1674.8274	1674.7936	0.0338	0	K.IVSDGNGMNAWVAWR.N
117 - 125	1045.5506	1044.5434	1044.5352	0.0081	0	K.GTDVQAWIR.G

Fig. 4.9 Mascot search results of coverage and matching peptide sequences obtained from the Mascot

zoomed-in spectrum of the region between 600 and 780 Da. Matching with theoretical digested peak list, peak of m/z 606.3 represents the N-terminal fragment of lysozyme (amino acids residue of 1–5: KVFGR). In comparison with the spectrum of native lysozyme in Fig. 4.10a, a new peak was observed in the spectrum of glycated lysozyme in Fig. 4.10b at m/z 768.4 with a mass difference of 162 Da, attributed to a glycation of the N-terminal fragment with one glucose moiety loaded to the peptide fragment with the dehydration of one molecule of water.

Meanwhile, in the mass region between m/z 1760 and 2000 in the MALDI-TOF-MS spectrum in Fig. 4.10b, only one dominant peak of m/z 1805.0 was observed, which was matched to a peptide fragment of lysozyme (amino acids residue 97–112: KIVSDGNGMNAWVAWR). In the corresponding region of the

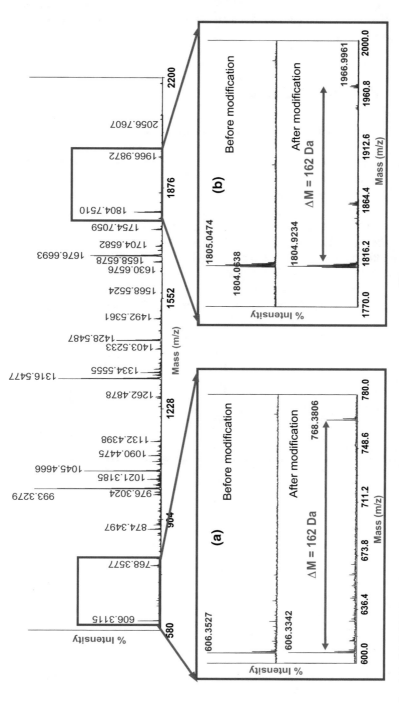

Fig. 4.10 MALDI-TOF-MS spectra of tryptically digested glucosylated lysozyme (LYZ-G^{50}-30) in the mass range *m/z* 600–780 (**a**) and *m/z* 1770–2000 (**b**) (the two potential glycated peptide peaks were labeled with mass shift of 162 Da)

glycated lysozyme, there was an additional peak at *m/z* 1967.0 in Fig. 4.10b. The peak at 1967.0 Da, with a mass increase of 162 Da, was the expected glucose molecule attached.

4.9.2   MALDI-TOF/TOF-MS

In order to obtain solid evidence of glycation and to identify the target amino acid, MALDI-TOF/TOF-MS was applied to the dominant peaks of digested lysozyme and new peaks of glycated peptide sequences. The precursor ions were isolated in the mass spectrometer and further collided at high energy in MALDI-TOF/TOF-MS to obtain the fragmentation ions. The product ions were mainly generated through peptide backbone cleavage (amide bond cleavage) under TOF/TOF experiments, and consecutive y-series ion could be observed. In the MALDI-TOF/TOF-MS spectra, fragment ions not only provide peptide sequence information and modification information, but also confirm the glycation sites of lysine in the modified lysozyme with mass difference of 162 Da.

In Fig. 4.11a, MS/MS spectrum of peak at *m/z* 606.3 is verified by y-series ions that the peptide sequence was KVFGR (residue: 1–5). Fragments in the spectrum of

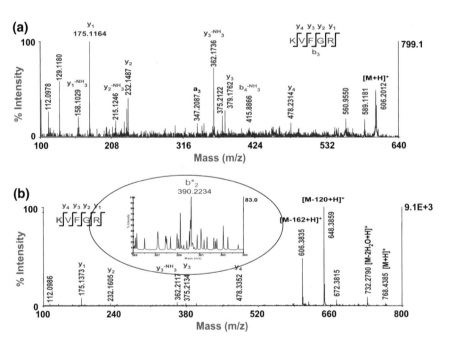

Fig. 4.11 MALDI-TOF/TOF-MS spectra of **a** peak at *m/z* 606.3 of protonated [KVFGR+H]⁺ peptide (amino acid: 1–5) and **b** peak at *m/z* 768.3 of protonated glucosylated [K*VFGR+H]⁺ peptide

the additional peak at m/z 768.4 ($[\mathbf{K}^*\text{VFGR+H}]^+$) displayed the same y-series ions at lower mass range and one weak modified b_2^*-ion at m/z 390.2, indicating that lysine 1 was the glycation site. The peaks at relatively high mass region (m/z 610–780 Da) were typical fragment ions of glucose by neutral loss of water (peak of m/z 732.3) and m/z 120 (peak of m/z 648.4), which occurred at the high collision energy where C–C bonds were broken [83, 84].

Using the same analytical method, the MALDI-TOF-TOF-MS spectrum of m/z 1804.7 ($[\text{KIVSDGNGMNAWVAWR+H}]^+$ (residue: 97–112)) and the matched additional peak at m/z 1966.9 ($[\mathbf{K}^*\text{IVSDGNGMNAWVAWR+H}]^+$) had almost the same y-series ions (Fig. 4.12), suggesting that lysine 97 was the glycation site in this peptide. Similarly, dominant y-ion peaks in MALDI-TOF/TOF-MS owing to free N terminus under higher collision energies also caused C–C bond cleavage. Lysine was the target glycated site in the Maillard reaction with sugars [47, 83, 85–87], and the MALDI-TOF/TOF-MS data show that the amino acid lysine 1 and lysine 97 of lysozyme were the glycation sites in this study.

4.9.3 Nano-LC-MS/MS

Although MALDI-TOF/TOF-MS analysis directly shows the modified peptides from mass differences and modification sites of the fragment ions, the sequence coverage of 55% was relatively low and only two glycation sites (lys1 and lys97) were verified in lysozyme. However, the ESI-MS spectra of intact glycated lysozyme clearly show the presence of four glucose moieties covalently bound to lysozyme. To solve the similar problem of low coverage in MALDI-TOF-MS peptide mapping, online LC-MS/MS analysis of the tryptic-digested glycated lysozyme was performed through reverse-phase nano-UPLC to improve the dynamic range for the detection of low-abundance ions. To further investigate the other two glycation sites, nano-RP-LC-MS/MS was employed. MS/MS data were first converted into the SEQUEST format and then searched against the SEQUEST algorithm using BioWorks 3.3, and two missed cleavages were set. After RP-LC separation, MS/MS data searching in Mascot provided near-complete coverage (95%) of the mixture of non-modified and modified lysozyme tryptic peptides.

The native peptide and the glycated peptides were fingerprinted in terms of their m/z values and then characterized from their product ion spectra. The MS/MS spectra of the two precursor ions at m/z 303.7134 of normal peptide KVFGR in Fig. 4.13a and m/z 384.7392 of matched glycated peptide $\mathbf{K}^*\text{VFGR}$ are shown in Fig. 4.13b, respectively. The two doubly protonated precursor ions both displayed the same y-ions (y3 and y4). MS/MS of $[\text{KVFGR+2H}]^{2+}$ shows the typical peptide fragmentation ion and NH_3 loss, and MS/MS of $[\mathbf{K}^*\text{VFGR+2H}]^{2+}$ also shows the classical fragmentation ions from the sugar moiety (neutral loss of water) and peptide moiety (y-ions). $[\mathbf{K}^*\text{VFGR+2H}]^{2+}$ was accompanied by abundant signals at m/z 366.7273 and 357.7190 corresponding to mass losses of 36 Da ($-2 \times \text{H}_2\text{O}$) and

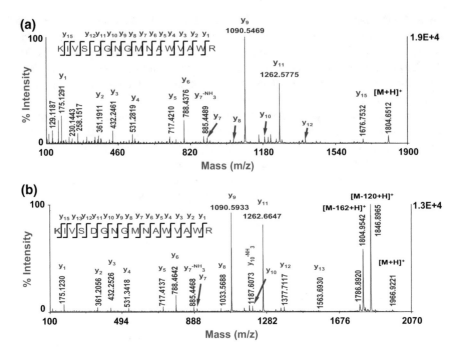

Fig. 4.12 MALDI-TOF/TOF-MS spectra of **a** peak at m/z 1804.7 corresponding to peptide [KIVSDGNGMNAWVAWR+H]$^+$ (amino acid: 97112) and **b** peak at m/z 1966.9 corresponding to glycated peptide [**K***IVSDGNGMNAWVAWR+H]$^+$

54 Da ($-3 \times H_2O$) by forming pyrylium, and the dominant peak was m/z 342.7164 by the loss of 84 Da ($-3 \times H_2O$–HCHO) at the sugar moiety forming furylium ions as shown in Fig. 4.13b, which was the typical fragment ion of the sugar moiety at low collision energy and confirmed the fragmentation mechanism proposed for ARPs in ESI-MS in previous chapters. The ion at m/z 303.7068 was the normal peptide after the loss of the whole sugar moiety MS/MS spectrum. The strong signal at m/z 294.1879 came from the 180 Da loss by theoretical calculation of a loss of the sugar moiety and one H_2O from the peptide (Fig. 4.14).

Besides the two glycated sites demonstrated by MALDI-TOF/TOF-MS, the other two glycated peptides were located and sequenced by LC-MS/MS. Figure 4.15 shows the product ion scan spectra of the precursor ions at m/z 525.28 and 606.31, corresponding to the doubly protonated native and glycated peptides [CELAAAMKR+2H]$^{2+}$ in Fig. 4.15a and [CELAAAM**K***R+2H]$^{2+}$ in Fig. 4.15b, respectively; the b-series and y-series ions in the product ions of the glycated species were comprehensive, in accordance with the results presented in the previous reports. Similarly, Fig. 4.16 shows the product ions of the precursor ions at m/z 667.35 and 748.38, which could be identified from their fragment ions as being [CKGTDVQAWIR+2H]$^{2+}$ and [C**K***GTDVQAWIR+2H]$^{2+}$, respectively. Fragmentation patterns arising from neutral losses of parts of the sugar adduct were

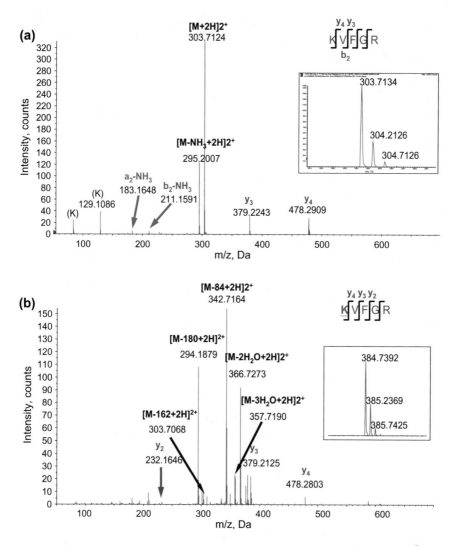

Fig. 4.13 MS/MS spectra of the ions at **a** *m/z* 303.7134 of normal peptide [KVFGR+2H]$^{2+}$ and **b** *m/z* 384.7392 of matched glycated peptide [KVFGR+Glc+2H]$^{2+}$

observed; the unchanged y_5-ion and the mass-increased b_5^{2+} fragment ions from the glycated peptide restricted the glycation site to within the first two residues (cysteine and lysine) from the N terminus. While not completely unequivocal, the product ions provide the confidence in locating the glycation site of lysine by restricting the number of possible residues.

All the four glycated peptides were identified by nano-LC-MS/MS, including the two peptides of **K*VFGR** and **K*IVSDGNGMNAWVAWR** characterized by MALDI-TOF/TOF. Besides [**K*IVSDGNGMNAWVAWR+3H**]$^{3+}$, the other three

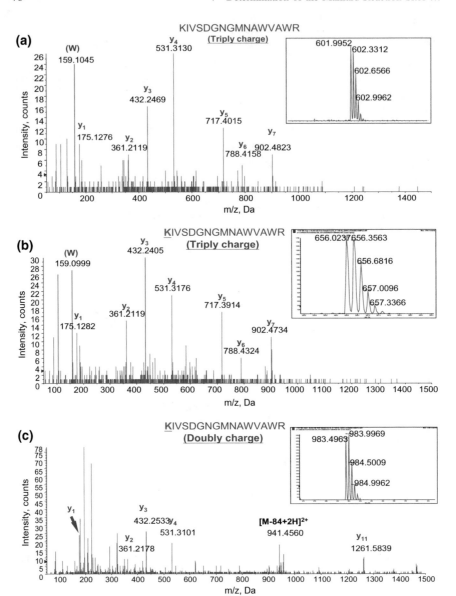

Fig. 4.14 MS/MS spectra of the ion at **a** *m/z* 601.9952 of normal peptide [KIVSDGNGMNAWVAWR+3H]³⁺, **b** *m/z* 656.0237 of matched glycated peptide [KIVSDGNGMNAWVAWR+Glc+3H]³⁺, and **c** *m/z* 983.9969 of matched glycated peptide [KIVSDGNGMNAWVAWR+Glc+2H]²⁺

glycol–peptides had typical neutral loss of 36, 54, 72, 84, and/or 162 Da, conforming the presence of glucose. Similarly, by the comparison of mass difference of precursor ions in the zoom scan, and fragmentation ions of the matched peptide

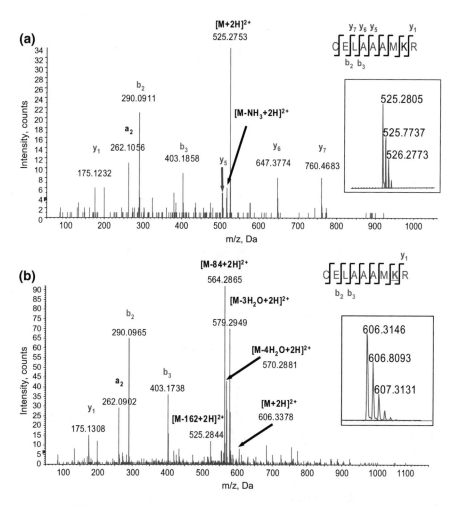

Fig. 4.15 MS/MS spectra of the ion at **a** m/z 525.2805 of normal peptide [CELAAAMKR+2H]$^{2+}$ and **b** m/z 606.3146 of matched glycated peptide [CELAAAMKR+Glc+2H]$^{2+}$

moiety and typical sugar moiety of glycopeptides, and also based on the previous studies, four glycation sites of lysozyme were verified by the LC-MS/MS and they were lys1 from the precursor ion of m/z 384.7392 [K*VFGR+2H]$^{2+}$ (Fig. 4.13), lys97 from the precursor ion of m/z 656.0237 of [K*IVSDGNGMNAWVAWR +3H]$^{3+}$ as shown in Fig. 4.13, lys13 from the precursor ion of m/z 606.3146 of [CELAAAMK*R+2H]$^{2+}$ as shown in Fig. 4.15, and lys116 from the precursor ion of m/z 748.3837 of [CK*GTDVQAWIR+2H]$^{2+}$ as shown in Fig. 4.16.

The MS/MS study described the MS-based monitoring of the Maillard reactions between glucose and lysozyme performed under controlled conditions. ESI-MS and MALDI-TOF-MS could monitor the Maillard reaction based on molecular weight changes, and spectra of the intact glycated lysozyme confirmed that at least four D-

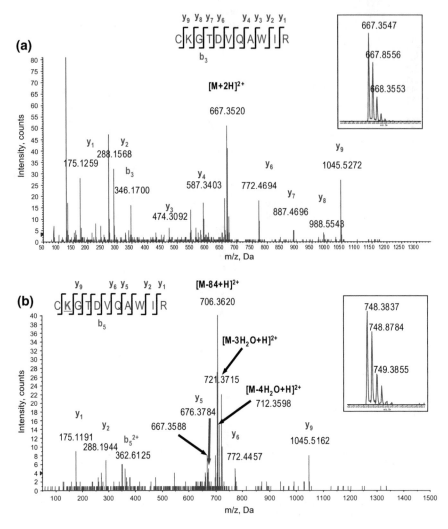

Fig. 4.16 MS/MS spectra of the ion at **a** *m/z* 667.3547 of normal peptide [CKGTDVQAWIR +2H]$^{2+}$ and **b** *m/z* 748.3837 of matched glycated peptide [CKGTDVQAWIR+Glc+2H]$^{2+}$

glucose reacted with the protein. The extent of glycosylation was first screened by observing the mass shifts in MS of the intact lysozyme, which influenced emulsifying properties. Thermal properties of glucosylated lysozyme were also evaluated by DSC.

MALDI-TOF/TOF-MS shows two glycation sites of lys1 and lys97. The other two glycated sites, lys13 and lys116, are detected by nano-LC-MS/MS indirectly. Analysis of the peptides of enzymatic digest of lysozyme revealed that glycation of protein led to modification of its conformation, resulting in the exposure of its active sites.

Compared with MALDI-TOF/TOF-MS, nano-LC-QqTOF-MS provided high coverage of the total digested protein after LC separation and more information of the sugar moiety of low energy collision technique. However, the method could not provide identification of the modified peptide residues (e.g., y*-, a*-, and b*-ions), and the peptide sequencing information obtained was poorer than that using MALDI-TOF/TOF-MS.

4.10 Summary

The study presented an systematical experiments on ARPs in the Maillard reaction between sugars and amino acids, peptides, and proteins. Such as, the effects of reactants, temperature, and time were elucidated. In brief, basic amino acids, such as lysine (N^{α}-acetyl-lysine), histidine, and arginine, preferentially reacted with reducing sugars such as glucose, maltose, and lactose in studied reaction conditions. Totally, in five selected sugars, the sugar effect on the reaction rate was as follows: glucose > lactose > maltose ≫ sucrose and raffinose. Reaction temperature played a key role on the Maillard reaction; the length of peptide slightly influences the reaction rate, whereas the position of lysine in the peptides had no discernible effects.

Aided by NRM structural identification and theoretical calculation of specific ARPs, the main gas-phase fragmentation pathways of ARPs in ESI-MS were proposed. The fragmentation behaviors of the ARPs in acid/peptide–sugar model showed two main dissociations of sugar moiety and peptide moiety. The fragmentation of sugar moiety occurred at relatively low collision energy and tended to take place preferentially by neutral loss of small molecules such as water and HCHO to form oxonium ion of $[M-n \times H_2O+H]^+$ ($n = 1-4$) and $[M-84+H]^+$ ($-3H_2O-HCHO$). The important fragmentation ion $[M-84+H]^+$ could be further dissociated at relative high collision energy. The spectra of MS^3 of $[M-84+H]^+$ not only displayed the peptide sequence, but also provided useful information about the glycation sites of the peptides with a*-, b*-, and y*-series ions and typical fragmentation ions of lysine residues.

Glucosylation influenced the emulsifying properties noticeably and significantly ($p < 0.05$). Both the values of EAI and the ESI* of lysozyme were increased by glucosylation to a certain degree, whereas EAI value was significantly ($p < 0.05$) decreased upon extended glucosylation. Thermal properties were also affected by glucosylation.

Potential reaction sites were obtained from the specific mass shift in peptide mapping of MALDI-TOF-MS and LC-ESI-MS. Further, these peptides were also isolated and the product ions were confirmed by MALDI-TOF/TOF and nano-LC-QqTOF-MS/MS. MALDI-TOF/TOF spectra provided glycation information of digested protein. With the combination of these two MS/MS techniques, all four glycation sites (Lys1, Lys13, Lys97, and Lys113) were identified unambiguously to specific residues of the tryptic peptides.

The methodology of MS-based techniques set up in this study should be helpful for analysis of food modification in future. The study represented a model of MRPs (ARPs) in three different systems were systematical studied using a wide variety of analytical and theoretical techniques. In the future, the study could be extended to other types MRPs. More theoretical calculation work could be designed to display the fragmentation behaviors of ARPs in comparison with instrumental results. In the characterization of glycation sites, the research could be extended to utilize multiple reaction monitoring (MRM), which might improve the detection of low-abundance glycated peptides. It is in our intention to extend the extent of glycation in protein–sugar systems and conduct experiments on analyzing more properties of modified food proteins.

References

1. Singh S, Gamlath S, Wakeling L (2007) Nutritional aspects of food extrusion: a review. Inter J Food Sci Tech 42(8):916–929
2. Ledl F, Schleicher E (1990) New aspects of the Maillard reaction in foods and in the human-body. Angew Chem Inter 29(6):565–594
3. Franzen KL, Kinsella JE (1976) Functional properties of succinylated and acetylated soy protein. J Agric Food Chem 24(4):788–795
4. Aoki T, Kitahata K, Fukumoto T et al (1997) Improvement of functional properties of beta-lactoglobulin by conjugation with glucose-6-phosphate through the Maillard reaction. Food Res Inter 30(6):401–406
5. Arntfield SD, Murray ED (1981) The influence of processing parameters on food protein functionality. 1. Differential scanning calorimetry as an indicator of protein denaturation. Can Food Sci Tech J 14(4):289–294
6. Kato A, Mifuru R, Matsudomi N et al (1992) Functional casein-polysaccharide conjugates prepared by controlled dry heating. Biosci Biotech Biochem 56(4):567–571
7. Kato A (2002) Industrial applications of Maillard-type protein-polysaccharide conjugates. Food Sci Tech Res 8(3):193–199
8. Oliver CM, Melton LD, Stanley RA (2006) Creating proteins with novel functionality via the Maillard reaction: a review. Cri Rev Food Sci Nut 46(4):337–350
9. Ajandouz EH, Tchiakpe LS, Dalle Ore F et al (2001) Effects of pH on caramelization and Maillard reaction kinetics in fructose-lysine model systems. J Food Sci 66(7):926–931
10. Ashoor SH, Zent JB (1984) Maillard browning of common amino-acids and sugars. J Food Sci 49(4):1206–1207
11. Wang JC, Kinsella JE (1976) Functional properties of alfalfa leaf protein—foaming. J Food Sci 41(3):498–501
12. Groubet R, Chobert JM, Haertle T (1999) Functional properties of milk proteins glycated in mild conditions. Sci Alimen 19(3–4):423–438
13. Darewicz M, Dziuba J (2001) The effect of glycosylation on emulsifying and structural properties of bovine beta-casein. Nahrung-Food 45(1):15–20
14. Achouri A, Boye JI, Yaylayan VA et al (2005) Functional properties of glycated soy 11S glycinin. J Food Sci 70(4):C269–C274
15. Herceg Z, Rezek A, Lelas V et al (2007) Effect of carbohydrates on the emulsifying, foaming and freezing properties of whey protein suspensions. J Food Eng 79(1):279–286
16. Nacka F, Chobert JM, Burova T et al (1998) Induction of new physicochemical and functional properties by the glycosylation of whey proteins. J Protein Chem 17(5):495–503

17. Van Der Veen M, Norde W, Stuart MC (2005) Effects of succinylation on the structure and thermostability of lysozyme. J Agric Food Chem 53(14):5702–5707
18. Van Teeffelen AMM, Broersen K, De Jongh HHJ (2005) Glucosylation of beta-lactoglobulin lowers the heat capacity change of unfolding; a unique way to affect protein thermodynamics. Protein Sci 14(8):2187–2194
19. Pham VT, Ewing E, Kaplan H et al (2008) Glycation improves the thermostability of trypsin and chymotrypsin. Biotech Bioeng 101(3):452–459
20. Broersen K, Voragen AGJ, Hamer RJ et al (2004) Glycoforms of beta-lactoglobulin with improved thermostability and preserved structural packing. Biotech Bioeng 86(1):78–87
21. Kato A, Minaki K, Kobayashi K (1993) Improvement of emulsifying properties of egg-white proteins by the attachment of polysaccharide through Maillard reaction in a dry state. J Agric Food Chem 41(4):540–543
22. Harwalkar VR, Ma CY (1987) Study of thermal-properties of oat globulin by differential scanning calorimetry. J Food Sci 52(2):394–398
23. Ma CY, Khanzada G, Harwalkar VR (1988) Thermal gelation of oat globulin. J Agric Food Chem 36(2):275–280
24. Raymond DE, Harwalkar VR, Ma CY (1992) Detection of incubator reject eggs by differential scanning calorimetry. Food Res Inter 25(1):31–35
25. Ma CY, Harwalkar VR (1996) Effects of medium and chemical modification on thermal characteristics of beta-lactoglobulin. J Therm Anal 47(5):1513–1525
26. Choi SM, Mine Y, Ma CY (2006) Characterization of heat-induced aggregates of globulin from common buckwheat (Fagopyrum Esculentum Moench). Inter J Biol Macromol 39(4–5):201–209
27. Koshiyama I, Hamano M, Fukushima D (1981) A heat denaturation study of the 11 s globulin in soybean seeds. Food Chem 6(4):309–322
28. Privalov PL (1982) Stability of proteins—proteins which do not present a single cooperative system. Adv Protein Chem 35:1–104
29. Somoza V (2005) Five years of research on health risks and benefits of Maillard reaction products: an update. Mol Nutr Food Res 49(7):663–672
30. Yamagishi SI, Ueda S, Okuda S (2007) Food-derived advanced glycation end products (AGEs): a novel therapeutic target for various disorders. Cur Pharm Design 13:2832–2836
31. Niwa T (2006) Mass spectrometry for the study of protein glycation in disease. Mass Spectrom Rev 25(5):713–723
32. Lapolla A, Fedele D, Seraglia R et al (2006) The role of mass spectrometry in the study of non-enzymatic protein glycation in diabetes: an update. Mass Spectrom Rev 25(5):775–797
33. Fay LB, Brevard H (2005) Contribution of mass spectrometry to the study of the Maillard reaction in food. Mass Spectrom Rev 24(4):487–507
34. Wenzl T, de la Calle MB, Anklam E (2003) Analytical methods for the determination of acrylamide in food products: a review. Food Addit Contam 20(10):885–902
35. Tagami U, Akashi S, Mizukoshi T et al (2000) Structural studies of the Maillard reaction products of a protein using ion trap mass spectrometry. J Mass Spectrom 35(2):131–138
36. Yeboah FK, Alli I, Yaylayan VA et al (2000) Monitoring glycation of lysozyme by electrospray ionization mass spectrometry. J Agric Food Chem 48(7):2766–2774
37. Smales CM, Pepper DS, James DC (2001) Evaluation of protein modification during anti-viral heat bioprocessing by electrospray ionization mass spectrometry. Rapid Commun Mass Spectrom 15(5):351–356
38. Lapolla A, Fedele D, Martano L et al (2001) Advanced glycation end products: a highly complex set of biologically relevant compounds detected by mass spectrometry. J Mass Spectrom 36(4):370–378
39. Roscic M, Versluis C, Kleinnijenhuis AJ et al (2001) The early glycation products of the Maillard reaction: mass spectrometric characterization of novel imidazolidinones derived from an opioid pentapeptide and glucose. Rapid Commun Mass Spectrom 15(12):1022–1029

40. Saraiva MA, Borges CM, Florencio MH (2006) Reactions of a modified lysine with aldehydic and diketonic dicarbonyl compounds: an electrospray mass spectrometry structure/activity study. J Mass Spectrom 41(2):216–228
41. Pearce KN, Kinsella JE (1978) Emulsifying properties of proteins—evaluation of a turbidimetric technique. J Agric Food Chem 26(3):716–723
42. Yeboah FK, Yaylayan VA (2001) Analysis of glycated proteins by mass spectrometric techniques: qualitative and quantitative aspects. Nahrung-Food 45(3):164–171
43. Meltretter J, Pischetsrieder M (2008) Application of mass spectrometry for the detection of glycation and oxidation products in milk proteins. Maillard React: Rec Adv Food Biomed Sci 1126:134–140
44. Corzo-Martinez M, Moreno FJ, Olano A et al (2008) Structural characterization of bovine beta-lactoglobulin-galactose/tagatose Maillard complexes by electrophoretic, chromato-graphic, and spectroscopic methods. J Agric Food Chem 56(11):4244–4252
45. Talasz H, Wasserer S, Puschendorf B (2002) Nonenzymatic glycation of histones in vitro and in vivo. J Cel Biochem 85(1):24–34
46. Kislinger T, Humeny A, Peich CC et al (2005) Analysis of protein glycation products by MALDI-TOF/MS. Maillard React 1043:249–259
47. Kislinger T, Humeny A, Seeber S et al (2002) Qualitative determination of early Maillard-products by MALDI-TOF mass spectrometry peptide mapping. Eur Food Res Tech 215(1):65–71
48. Mangino ME (1994) Protein interactions in emulsions: protein-lipid interactions. Protein Func Food Sys 9:147–179
49. Mohanty B, Mulvihill DM, Fox PF (1988) Emulsifying and foaming properties of acidic caseins and sodium caseinate. Food Chem 28(1):17–30
50. Darewicz M, Dziuba J, Mioduszewska H et al (1999) Modulation of physico-chemical properties of bovine beta-casein by nonenzymatic glycation associated with enzymatic dephosphorylation. Acta Aliment 28(4):339–354
51. Caessens P, Visser S, Gruppen H et al (1999) Emulsion and foam properties of plasmin derived beta-casein peptides. Inter Dairy J 9(3–6):347–351
52. Saeki H (1997) Preparation of neoglycoprotein from carp myofibrillar protein by Maillard reaction with glucose: biochemical properties and emulsifying properties. J Agric and Food Chem 45(3):680–684
53. Kato A, Kobayashi K (1991) Excellent emulsifying properties of protein dextran conjugates. Acs Sympos Series 448:213–229
54. Nakamura S, Kato A, Kobayashi K (1992) Bifunctional lysozyme galactomannan conjugate having excellent emulsifying properties and bactericidal effect. J Agric Food Chem 40 (5):735–739
55. Kato A, Sasaki Y, Furuta R et al (1990) Functional protein polysaccharide conjugate prepared by controlled dry-heating of ovalbumin dextran mixtures. Agric Biol Chem 54(1):107–112
56. Moreno FJ, Lopez-Fandino R, Olano A (2002) Characterization and functional properties of lactosyl caseinomacropeptide conjugates. J Agric Food Chem 50(18):5179–5184
57. Xu H, Lekkerkerker HNW, Baus M (1992) Nematic-smectic-a and nematic-solid transitions of parallel hard spherocylinders from density functional theory. Europhys Lett 17(2):163–168
58. Vroege GJ, Lekkerkerker HNW (1992) Phase-transitions in lyotropic colloidal and polymer liquid-crystals. Prog Phys 55(8):1241–1309
59. Cao X, Wen YF, Zhang SC et al (2006) A heat-resistant emulsifying sizing agent for carbon fibers. New Carbon Mate 21(4):337–342
60. Cao Y, Xiong J, Yuan J Hydroxy camptothecin emulsion and its preparation method. Patent, CN1493289; CN1254245
61. Cao YH, Dickinson E, Wedlock DJ (1991) Influence of polysaccharides on the creaming of casein-stabilized emulsions. Food Hydrocoll 5(5):443–454
62. Dickinson E, Ma JG, Povey MJW (1994) Creaming of concentrated oil-in-water emulsions containing xanthan. Food Hydrocoll 8(5):481–497

63. Dickinson E, Pawlowsky K (1996) Rheology as a probe of protein-polysaccharide interactions in oil-in-water emulsions. Gums Stab Food Indust 8:181–191
64. KobersteinHajda A, Dickinson E (1996) Stability of water-in-oil-in-water emulsions containing faba bean proteins. Food Hydrocoll 10(2):251–254
65. Kim HJ, Choi SJ, Shin WS et al (2003) Emulsifying properties of bovine serum albumin-galactomannan conjugates. J Agric Food Chemi 51(4):1049–1056
66. Dickinson E, Galazka VB, Anderson DMW (1991) Emulsifying behavior of gum-arabic. 1. Effect of the nature of the oil phase on the emulsion droplet-size distribution. Carbohydr Polym 14(4):373–383
67. Dickinson E, Galazka VB, Anderson DMW (1991) Emulsifying behavior of gum-arabic. 2. Effect of the gum molecular-weight on the emulsion droplet-size distribution. Carbohydr Polym 14(4):385–392
68. Nagasawa K, Ohgata K, Takahashi K et al (1996) Role of the polysaccharide content and net charge on the emulsifying properties of beta-lactoglobulin-carboxymethyldextran conjugates. J Agric Food Chem 44(9):2538–2543
69. Nagasawa K, Takahashi K, Hattori M (1996) Improved emulsifying properties of beta-lactoglobulin by conjugating with carboxymethyl dextran. Food Hydrocoll 10(1):63–67
70. Yadav S, Ahmad F (2000) A new method for the determination of stability parameters of proteins from their heat-induced denaturation curves. Anal Biochem 283(2):207–213
71. Kumita JR, Johnson RJK, Alcocer MJC et al (2006) Impact of the native-state stability of human lysozyme variants on protein secretion by Pichia pastoris. FEBS J 273(4):711–720
72. Wright DJ (1984) Thermoanalytical methods in food research. Cri Rep Appl Chem 5:1–36
73. Li-Chan ECY, Ma CY (2002) Thermal analysis of flaxseed (Linum usitatissimum) proteins by differential scanning calorimetry. Food Chem 77(4):495–502
74. Boye JI, Ma CY, Ismail A et al (1997) Molecular and microstructural studies of thermal denaturation and gelation of beta-lactoglobulins A and B. J Agric Food Chem 45(5): 1608–1618
75. Boye JI, Ma CY, Ismail A (2004) Thermal stability of beta-lactoglobulins A and B: effect of SDS, urea, cysteine and N-ethylmaleimide. J of Dairy Res 71(2):207–215
76. Sola RJ, Al-Azzam W, Griebenow K (2006) Engineering of protein thermodynamic, kinetic, and colloidal stability: Chemical glycosylation with monofunctionally activated glycan. Biotech Bioeng 94(6):1072–1079
77. Sola RJ, Griebenow K (2006) Chemical glycosylation: new insights on the interrelation between protein structural mobility, thermodynamic stability, and catalysis. FEBS Lett 580 (6):1685–1690
78. Kato Y, Watanabe K, Sato Y (1981) Effect of Maillard reaction on some physical-properties of ovalbumin. J Food Sci 46(6):1835–1839
79. Yeargans GS, Seidler NW (2003) Carnosine promotes the heat denaturation of glycated protein. Biochem Biophys Res Commun 300(1):75–80
80. Choi SM, Ma CY (2005) Conformational study of globulin from common buckwheat (Fagopyrum esculentum Moench) by Fourier transform infrared spectroscopy and differential scanning calorimetry. J Agric Food Chem 53(20):8046–8053
81. Berman HM, Battistuz T, Bhat TN et al (2002) The protein data bank. Acta Crystallogr Sect D 58:899–907
82. Berman HM, Westbrook J, Zardecki C et al (2003) The protein data bank. Protein Struct 389–405
83. Brancia FL, Bereszczak JZ, Lapolla A et al (2006) Comprehensive analysis of glycated human serum albumin tryptic peptides by off-line liquid chromatography followed by MALDI analysis on a time-of-flight/curved field reflectron tandem mass spectrometer. J Mass Spectrom 41(9):1179–1185
84. Fenaille F, Parisod V, Tabet JC et al (2005) Carbonylation of milk powder proteins as a consequence of processing conditions. Proteom 5(12):3097–3104

85. Fenaille F, Morgan F, Parisod V et al (2004) Solid-state glycation of beta-lactoglobulin by lactose and galactose: localization of the modified amino acids using mass spectrometric techniques. J Mass Spectrom 39(1):16–28
86. Ortwerth BJ, Slight SH, Prabhakaram M et al (1992) Site-specific glycation of lens crystallins by ascorbic-acid. Biochim Biophys Acta 1117(2):207–215
87. Tressi R, Piechotta CT, Rewicki D et al (2002) Modification of peptide lysine during Maillard reaction of D-glucose and D-lactose. Maillard React 1245:203–209

Printed in the United States
By Bookmasters